Visibility and Work Zone Traffic Control 2018

Transportation Research Record: Journal of the Transportation Research Board

Volume 2672, Issue 16

ISBN: 978-1-5108-7734-4

Printed from e-media with permission by:

Curran Associates, Inc.
57 Morehouse Lane
Red Hook, NY 12571

Some format issues inherent in the e-media version may also appear in this print version.

Published by Sage Publications USA

ISBN (Print) 978-1-5108-7734-4
ISBN 2018 Printed Set (All Issues) 978-1-5108-7735-1

Printed by Curran Associates, Inc. (2018)

Additional copies of this publication are available from:

Curran Associates, Inc.
57 Morehouse Lane
Red Hook, NY 12571 USA
Phone: 845-758-0400
Fax: 845-758-2633
Email: curran@proceedings.com
Web: www.proceedings.com

Volume 2672 Number 16 December 2018

TRR
JOURNAL OF THE TRANSPORTATION RESEARCH BOARD

Contents

Visibility and Work Zone Traffic Control 2018

Articles

TRR
JOURNAL OF THE TRANSPORTATION RESEARCH BOARD

Article

Transportation Research Record
2018, Vol. 2672(16) 1–13

DOI: 10.1177/0361198118758050
journals.sagepub.com/home/trr
SAGE

Application of LiDAR and Connected Vehicle Data to Evaluate the Impact of Work Zone Geometry on Freeway Traffic Operations

Michelle M. Mekker[1], Yun-Jou Lin[1], Magdy K. I. Elbahnasawy[1], Tamer S. A. Shamseldin[1], Howell Li[1], Ayman F. Habib[1], and Darcy M. Bullock[1]

Abstract

Extensive literature exists regarding recommendations for lane widths, merging tapers, and work zone geometry to provide safe and efficient traffic operations. However, it is often infeasible or unsafe for inspectors to check these geometric features in a freeway work zone. This paper discusses the integration of LiDAR (Light Detection And Ranging)-generated geometric data with connected vehicle speed data to evaluate the impact of work zone geometry on traffic operations. Connected vehicle speed data can be used at both a system-wide (statewide) or segment-level view to identify periods of congestion and queueing. Examples of regional trends, localized incidents, and recurring bottlenecks are shown in the data in this paper. A LiDAR-mounted vehicle was deployed to a variety of work zones where recurring bottlenecks were identified to collect geometric data. In total, 350 directional miles were covered, resulting in approximately 360 GB of data. Two case studies, where geometric anomalies were identified, are discussed in this paper: a short segment with a narrow lane width of 10–10.5 feet and a merging taper that was about 200 feet shorter than recommended by the Manual on Uniform Traffic Control Devices. In both case studies, these work zone features did not conform to project specifications but were difficult to assess safely by an inspector in the field because of the high volume of traffic. The paper concludes by recommending the use of connected vehicle data to systematically identify work zones with recurring congestion and the use of LiDAR to assess work zone geometrics.

Monitoring traffic and maintaining roadway capacity is an important task for transportation agencies. In work zones on high-speed facilities this is especially important, as queueing can be unexpected to drivers. This paper discusses the integration of connected vehicle traffic data with geometric data from mobile LiDAR (Light Detection And Ranging) to identify and diagnose recurring capacity problems in freeway work zones. Figure 1 is a photo of interstate queueing upstream of a work zone on a rural section of I-70 in Indiana in Vigo County. This work zone is called the I-70 Vigo work zone throughout the rest of this paper. A map of the I-70 Vigo work zone is shown in Figure 2*a*.

There are a variety of data sources and collection methods available for traffic monitoring. These data sources can be useful in a range of settings from system-wide (Figure 2*b*), district (Figure 2*c*), route (Figure 2*d*) and individual work zone location (Figure 2*e*). In Indiana, connected vehicle speed data is used to monitor traffic congestion statewide and in specific corridors, such as work zones. Although the Indiana Department of Transportation (INDOT) and other agencies have tools for identifying recurring work zone congestion, it can be challenging to identify geometric conditions that lead to unexpected capacity reduction and subsequent queueing (Figure 1). The following section briefly summarizes the literature, followed by examples of how connected vehicle data and LiDAR can be used to identify underlying geometric conditions that can negatively affect work zone capacity and lead to queueing. Two examples, a narrow lane width and a short merging taper, are presented to illustrate how connected vehicle and LiDAR data can be used in an integrated manner to improve agency management of construction work zones.

[1]Purdue University, West Lafayette, IN

Corresponding Author:
Address correspondence to Darcy M. Bullock: darcy@purdue.edu

Figure 1. Queued traffic upstream of an Indiana work zone.

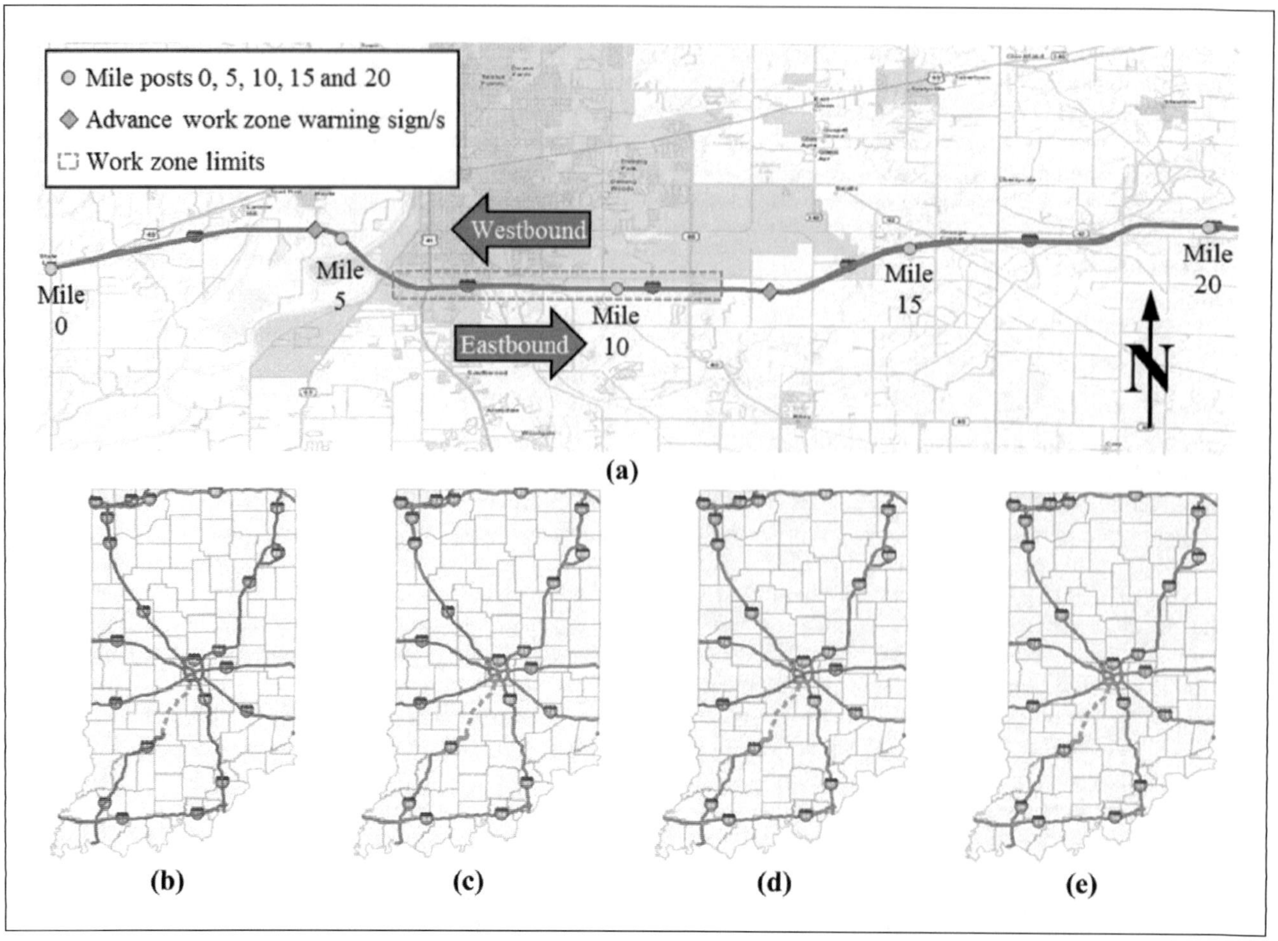

Figure 2. Selected interstate segments with travel speeds characterized by probe data: (a) close-up map of I-70 Vigo work zone, mile post 0 to 20; (b) statewide; (c) Crawfordsville District; (d) I-70; (e) I-70 Vigo work zone.

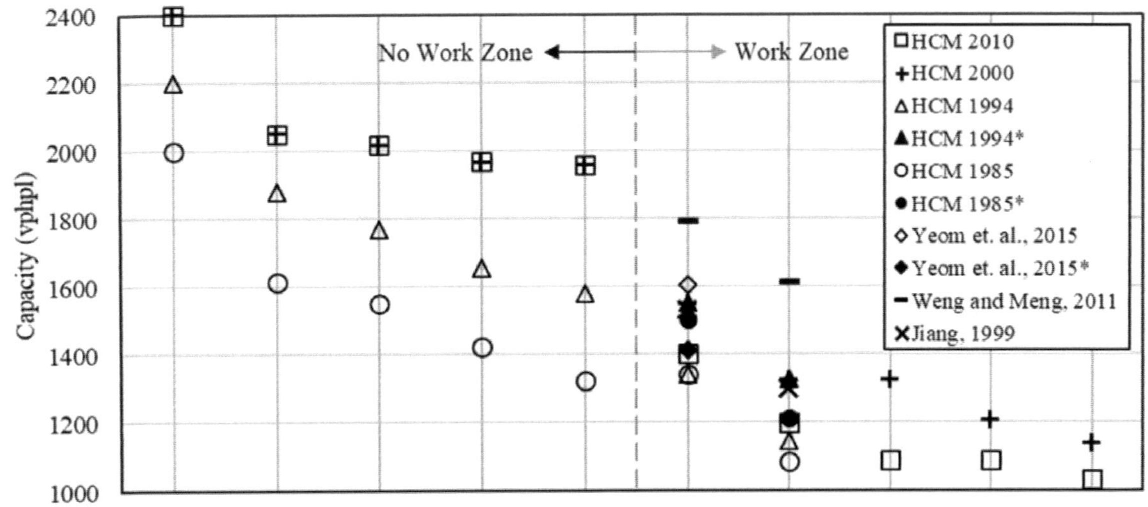

Lane Width (ft)	12	12	11	10	10	12	12	11	10	9
Shoulder Width (ft)	6	6	4	2	0	N/A	N/A	N/A	N/A	N/A
Percent Trucks (%)	0	34	34	34	34	0	34	34	34	34

*Concrete barriers present.

Figure 3. Summary of freeway capacity estimation for different scenarios.

Literature Review

Many agencies have policies in place to maintain capacity and reduce congestion from work zones on their roadways (*1*), which are important in the contract letting and design phases of road construction. Equally important is the administration and enforcement of those policies. Rouphail, Yang, and Fazio found that there were significant "discrepancies between standards and practice," and that there were higher speed variations in work zones where these discrepancies existed in their study of short- and long-term work zones (*2*). To manage work zone traffic control compliance and quality, the New York State Department of Transportation developed a Construction Health and Safety Program (*3*), which includes a quality assurance program and inspection procedures (*4*). Gambatese and Johnson conducted a study of 39 work zones over 5 years to determine what affected the quality, consistency, and safety of traffic control plans before and during construction (*5*).

Many agency policies and design manuals require or recommend that minimum lane widths and taper lengths be maintained in work zones. There is extensive literature regarding the estimation of capacity with lane width as a factor. Figure 3 summarizes the capacity estimation results for ten scenarios using methods from various Highway Capacity Manuals (HCMs) (*6–8*) and only a subset of the literature on the topic (*9–11*). For the work zone scenarios, the shoulder widths are not applicable because the capacity estimation models did not include shoulder width.

There is sparse literature regarding the effects of taper length on capacity. Sharp and Harwood studied the effects of different taper lengths and design speeds on traffic behavior at four sites in the 1970s (*12*). Theiss, Finley, and Ullman investigated the effects of reduced taper lengths only on lower-speed urban arterials (*13*). Most studies regarding taper length refer to the Manual on Uniform Traffic Control Design (MUTCD) (*14*). INDOT uses values for minimum required taper length similar to the MUTCD but rounds the values up to easily convert them to the equivalent number of broken line segments (*15*). Assuming that the skip lines are of the correct length and spacing, construction workers can count the number of skip lines between the start and end of the taper.

LiDAR systems onboard terrestrial mobile platforms have emerged as a prominent tool for collecting high-density point clouds along object surfaces relative to a global reference frame (*16,17*). The use of LiDAR in the context of transportation is a quickly expanding topic in literature. In 2013, National Cooperative Highway Research Program

Report 748 was published, which included guidelines for using mobile LiDAR in transportation applications (*18*). Chang et al. also published guidance on how agencies could practically use and deploy LiDAR in 2014 (*19*). Chang et al. discuss common transportation applications and acquisition options of LiDAR. For example, the authors recommend the mobile LiDAR platform for applications such as construction clearance measurement, corridor mapping, safety assessments, and traffic operations.

Data Sources

In this study, two data sources were integrated to identify and diagnose recurring capacity problems in work zones: speed data collected from connected vehicles and geospatial data collected via mobile LiDAR. As discussed briefly in the previous section, there is extensive literature regarding the use of traffic data and LiDAR data separately. However, there is sparse literature on the integration of connected traffic data and LiDAR. This section discusses these two data sources.

Speed Data from Connected Vehicles

These data are collected by a third-party vendor from various sources in real time, including but not limited to: freight, in-vehicle GPS units, and smart mobile devices. These data, coming from 1–2% of vehicles on the roadway, is then aggregated into a space mean speed every minute for segments that are an average 0.88 miles in length. In Indiana, there are approximately 2600 segments covering all of the 2250 directional miles of interstate.

These speed data are used in Indiana for both real-time traffic monitoring and historical analysis. A performance measure that is commonly used is mile-hours by speed bin (*20*) (Figure 4). This measure is both spatially and temporally weighted and is the product of the duration of time that a specified condition persists in a segment and the length of that segment. In Figure 4, *a* and *b*, the vertical gray lines represent the boundaries between segments (defined by the vendor) and the dotted orange lines represent the work zone boundaries defined by INDOT. Each column between the segment boundaries represents the duration of time in which that segment had an average speed within the speed bins (see color scale for Figure 4). The area under the curves in Figure 4, *a* and *b*, is the total mile-hours, which can be calculated using Equation 1. These graphs in particular are used to highlight areas and time periods of congestion. Figure 4*c* displays the summation of mile-hours by hour of the day. Figure 4*d* is the summation of mile-hours for April 14, 2017, for both directions in the work zone. Agencies may have different perspectives on the most appropriate speed threshold, but the most commonly used speed threshold in Indiana to screen for congestion is 45 mph, although other thresholds, such as 30 or 15 mph, may be appropriate. A sample calculation for the 45 mph threshold is shown below.

$$MH_{45} = \sum_{i=1}^{n} (L_i \times t_i) \qquad (1)$$

Where:

MH_{45} = total mile-hours of operation < 45 mph

n = total number of segments in selected route

L_i = length of segment i in miles

t_i = duration of time in hours where segment i as an average speed < 45 mph

Example Calculation:

Time Range: 4/14/2017 00:00 EDT – 4/14/2017 23:59 EDT

Location: I-70, mile post 0 to 18

I-70 W (Figure 4*a*):

$$\begin{aligned} MH_{45,W} = {} & (1.15mi \times 0.00hr) + (0.27mi \times 0.00hr) + \\ & (1.12mi \times 0.00hr) + (1.14mi \times 0.02hr) + \\ & (1.02mi \times 0.00hr) + (0.41mi \times 0.00hr) + \\ & (1.02mi \times 0.00hr) + (1.10mi \times 0.25hr) + \\ & (0.27mi \times 0.02hr) + (1.04mi \times 0.00hr) + \\ & (1.06mi \times 0.00hr) + (1.10mi \times 0.00hr) + \\ & (1.12mi \times 0.00hr) + (1.03mi \times 0.00hr) + \\ & (1.03mi \times 0.00hr) = 3.53 \text{ mile-hours} \end{aligned}$$

I-70 E (Figure 4*b*):

$$\begin{aligned} MH_{45,E} = {} & (0.90mi \times 0.00hr) + (1.06mi \times 0.00hr) + \\ & (1.26mi \times 0.00hr) + (0.47mi \times 0.00hr) + \\ & (1.12mi \times 0.00hr) + (1.10mi \times 0.00hr) + \\ & (0.70mi \times 0.20hr) + (0.41mi \times 1.82hr) + \\ & (1.12mi \times 3.92hr) + (1.05mi \times 3.73hr) + \\ & (1.02mi \times 0.00hr) + (1.11mi \times 0.00hr) + \\ & (1.03mi \times 0.00hr) + (1.11mi \times 0.00hr) + \\ & (1.03mi \times 0.00hr) + (1.04mi \times 0.00hr) \\ & = 9.19 \text{ mile-hours} \end{aligned}$$

Total (Figure 4*d*): $MH_{45} = MH_{45,W} + MH_{45,E} = 3.53 + 9.19 = 12.72$ mile-hours

The mile-hours performance measure can be used from the system-wide level to the corridor/work zone level to identify recurring congestion, changing traffic patterns, and outlier events. Figure 5*a* is a plot of the mile-hours of congestion by day on all interstates in Indiana. Callout "i" is the date when a winter storm swept across the northern half of the state and left roadways covered in ice that dramatically reduced speeds. This winter storm is also visible when viewing the INDOT Crawfordsville District (Figure 5*b*) and I-70 (Figure 5*c*). This winter storm did not have as large of an

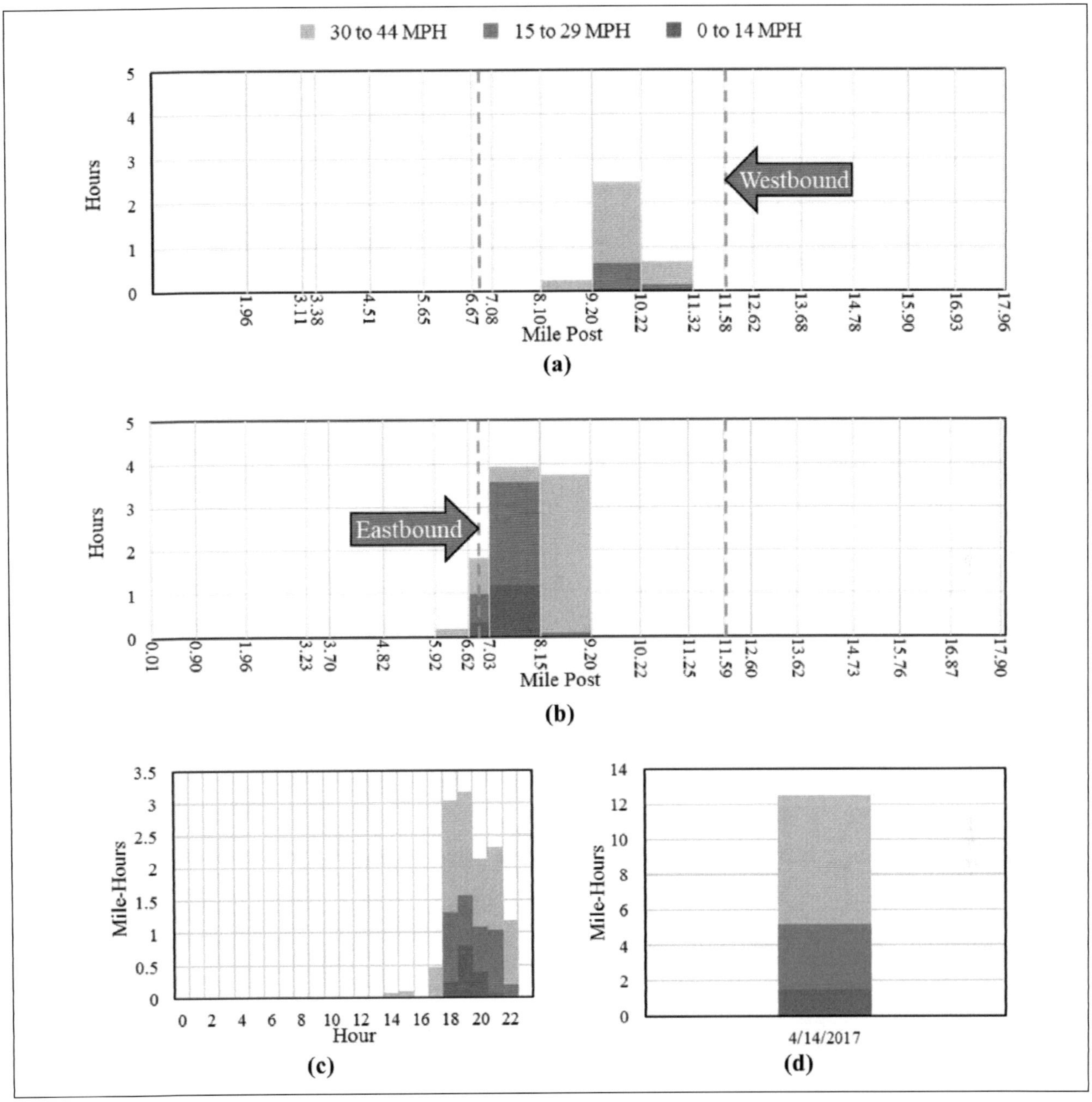

Figure 4. Operating speed characteristics of I-70, mile post 0 to 18, on April 14, 2017: (*a*) I-70 Westbound speed profile (hours operating in certain speed bin by mile post); (*b*) I-70 Eastbound speed profile (hours operating in certain speed bin by mile post); (*c*) mile-hours by speed bin by hour of day; (*d*) mile-hours by speed bin.

impact in the I-70 Vigo work zone (Figure 5*d*). However, Thanksgiving week travel (callout "ii"), a different winter storm on January 30 (callout "iii"), and increased congestion at the resumption of construction in April (callout "iv") are notable when examining the Vigo work zone segment of I-70 (Figure 5*d*). The next section of the paper describes the geometric data that are used to identify lane widths over extended road sections to determine if there are any sections that do not conform to project specifications or standards.

Geometric Data from LiDAR

A LiDAR system includes a laser ranging and scanning unit. For the laser ranging, the LiDAR system emits a pulse to estimate the distance based on the transmission time between the firing point and its footprint. Moreover, every pulse has an intensity defined by the return strength of the laser pulse. The intensity can indicate the reflectivity of an object hit by the laser pulse. For the scanning mechanism, laser scanners

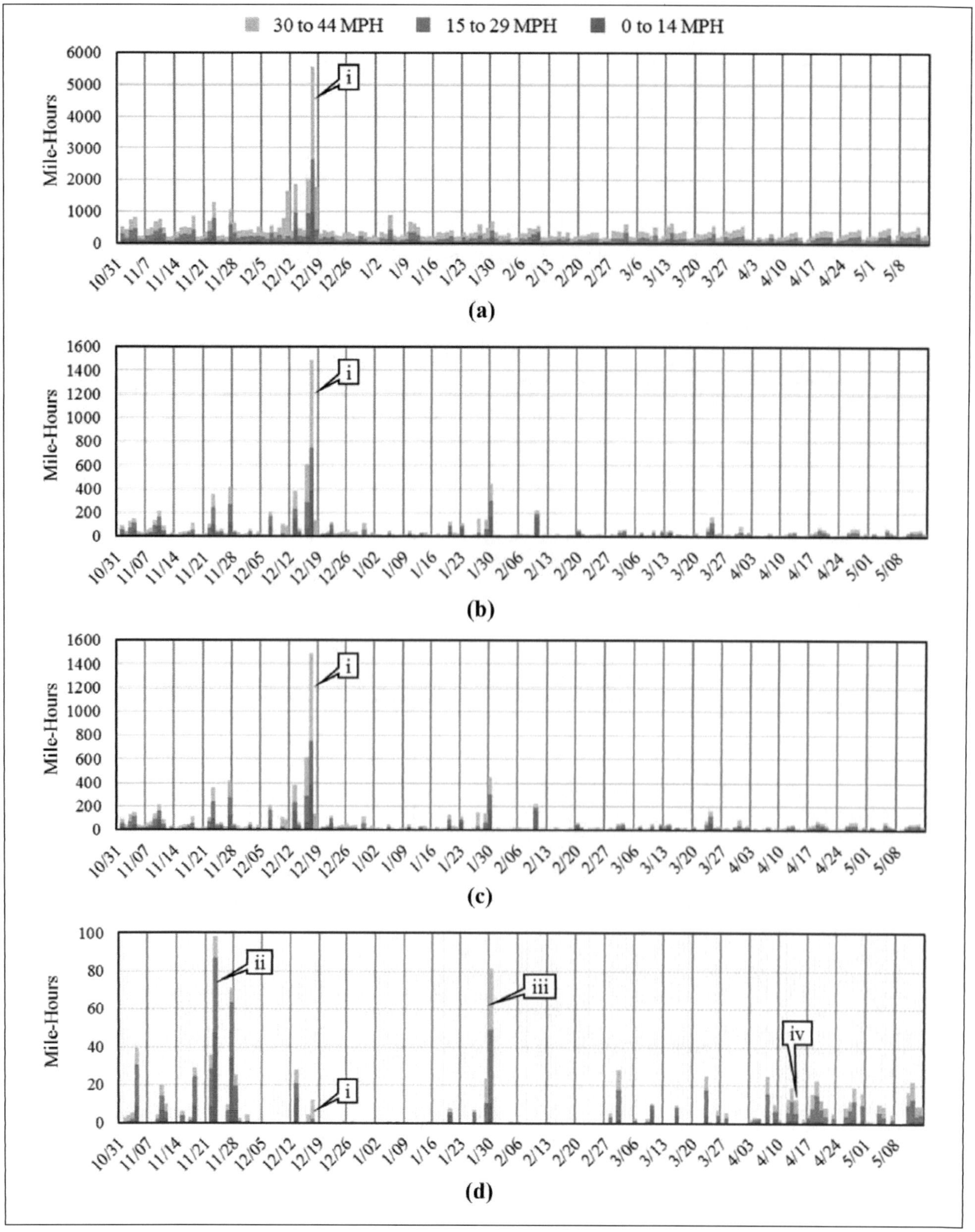

Figure 5. Mile-hours of congestion from statewide to work zone level: (*a*) statewide speed profile (routes shown as red in Figure 2*b*); (*b*) Crawfordsville District (routes shown as red in Figure 2*c*); (*c*) I-70 (routes shown as red in Figure 2*d*); (*d*) I-70 Vigo work zone (routes shown as red in Figure 2*e*).

Figure 6. Illustration of point positioning of a directly geo-referenced LiDAR system.

can be mainly classified into two categories: single laser scanners steered by a mirror and used for static scanning, and rotating multi-beam laser scanners used for terrestrial mobile mapping. A typical directly geo-referenced LiDAR system consists of a laser scanner, Global Navigation Satellite System (GNSS), and Inertial Navigation System (INS), which can provide accurate position and orientation of the vehicle platform.

A directly geo-referenced LiDAR system comprises three coordinate systems (mapping frame, m, GNSS/INS body frame, b, and laser unit frame, Lu) as illustrated in Figure 6. These coordinate systems and their spatial/rotational relationships are used to define the mapping coordinates (callout "i") of a given point, P, acquired from a mobile LiDAR mapping system, as given in Equation 2. The coordinates of point P relative to the laser unit coordinate system can be defined as $r_P^{Lu}(t)$ (callout "ii") by Equation 3, where α is the vertical angle determined by the fired laser beam ID; β is the horizontal angle, which depends on the rotation of the laser unit; ρ is the range defined by the distance from firing point to the footprint of the laser beam; and t is the time. The lever arm angle, r_{Lu}^b, and boresight angle, R_{Lu}^b, (callout "iii") between the laser unit and body frame coordinate systems are time-independent since the laser scanner and body frame are rigidly fixed relative to each other. The lever arm and boresight angles can be derived from a system calibration using conjugate targets in multiple drive runs (*21–23*). The GNSS/INS integration provides the time-dependent position, $r_b^m(t)$, and orientation, $R_b^m(t)$, (callout "iv") relating the mapping frame to the body frame.

$$r_P^m = r_b^m(t) + R_b^m(t) r_{Lu}^b + R_b^m(t) R_{Lu}^b r_P^{Lu}(t) \tag{2}$$

$$r_P^{Lu}(t) = \begin{pmatrix} x \\ y \\ z \end{pmatrix} = \begin{pmatrix} \rho(t)\cos\beta(t)\cos\alpha(t) \\ \rho(t)\cos\beta(t)\sin\alpha(t) \\ \rho(t)\sin\beta(t) \end{pmatrix} \tag{3}$$

After reconstructing the geo-referenced point cloud, features of interest can be identified and extracted automatically. Objects that are highly reflective can be identified using the intensity data in a given point cloud. Figure 7 shows some features of interest in the I-70 Vigo work zone in three different images: a camera view of the mobile LiDAR from behind, a camera view from the vehicle with the mobile LiDAR, and the point cloud at that location. Callouts "i," "ii," and "iii" are channelizing devices with reflective coatings or tapes. Callouts "iv" and "v" are the right and center lane markings. The lane widths along this route can be evaluated from the high-intensity points that represent lane markings. Callouts "vi" and "vii" in Figure 7*a* correspond to the on-board camera and the LiDAR unit. A great deal of research has been developed regarding the use of LiDAR for the recognition of the objects along roads (*24,25*), the improvement of road safety (*26*), automated driving (*27*), and risk management (*28*).

Figure 7. Example LiDAR data near mile post 11.5 in I-70 Vigo WB work zone: (*a*) LiDAR-mounted vehicle; (*b*) camera-view at vehicle location (callout "vi" in Figure 7*a*); (*c*) point cloud from LiDAR at vehicle location (callout "vii" in Figure 7*a*).

Deployment

Between August 2016 and June 2017, LiDAR data were collected on ten separate occasions and covering over 350 directional miles. Throughout Indiana, 18 work zones were being monitored using the connected vehicle speed data during this time period. The locations of the LiDAR data collections were selected from these work zones based on the most recent weeks of speed data. Potential sites were selected when there was recurring congestion or crashes at a consistent location, such as that seen in Figure 5*d*. Two case studies were selected for detailed discussion on how connected vehicle data can be used to identify work zones with significant congestion, and how LiDAR can be used to identify geometric conditions that deviate from designs and are likely contributing to the work zone queueing.

First, all agencies apply highly reflective glass beads or elements to pavement markings to improve night visibility. Well-maintained retro-reflective markings provide very high contrast in LiDAR point clouds (Figure 7*c*) and can be easily extracted via automated processes. However, extraction requires manual processing if markings are old and have poor retro-reflectivity characteristics.

Second, environmental conditions need to be considered when using LiDAR. Unlike video data collection, LiDAR can be used at night. In one nighttime lane closure, workers in reflective gear standing in front of a flood light were more visible in the LiDAR point cloud than could be visually observed by the driver. However, LiDAR performance significantly degrades when there is precipitation.

Third, the speed at which the mobile LiDAR travels affects the density of the data point cloud. Slower speeds will produce denser point clouds but may be less safe on high-speed facilities, such as freeways. In this study, 40 mph was chosen because it was the minimum allowable speed on Indiana interstates. All data collection runs were conducted with two vehicles: the vehicle mounted with the mobile LiDAR and a shadow vehicle equipped with strobe lights (not pictured in this paper) to alert other drivers to the slow-moving vehicles. As LiDAR technology matures and sampling frequency increases, these speed constraints will not be a concern. In fact, there are currently LiDAR systems that can be deployed at 70 mph that will provide sufficient resolution, but their cost exceeded the budget of this study.

Lastly, the LiDAR data is reproducible. LiDAR data was collected in the I-70 Vigo work zone twice and on dates 6 months apart (November 5, 2016, and May 2, 2017). Aside from the movement of drums for a lane closure, the lane markings and concrete barriers in this work zone remained the same. For the 20-mile data set in the westbound direction, the root mean square error of the lane width measurements were 1.6 cm (0.6 in).

Connecting Speed Data to Geometric Features

Two case studies where demand exceeded capacity for portions of the day are discussed. It was hypothesized that geometric features introduced some unanticipated capacity reductions in the work zone. These features were not identifiable by eye, and were in locations where it would have been unsafe for an inspector to manually measure. These case studies demonstrate how inspectors can correlate traffic impacts to specific work zone geometry, which can potentially assist in holding contractors accountable and improving both safety and throughput in work zones.

Case Study 1: Lane Width

The first case study is from the I-70 Vigo work zone (Figure 2*a*, Figure 4, and Figure 5*d*). This location has an annual average daily traffic (AADT) figure of 25,526, with 46% traveling westbound and 53% commercial vehicles (FHWA class 4 and up). In the westbound direction, there is a segment from mile post 12 to 10 that shifts the two lanes of travel toward the left shoulder and there is a concrete barrier on the right side of the roadway (Figure 8*a*). On this corridor, commercial vehicles make up about 40% of traffic volumes. At mile post 11.62, there is a brief section where the lane width decreases to 10.01 ft (Figure 8*b* and *c*). At this specific location, the distance from the edge line to the concrete barrier is 1.48 ft. In nearby locations, the distance to the barrier decreases to less than 1 ft. For this work zone, the minimum lane width allowed by INDOT is 11 ft. Figure 8*d* shows a longitudinal plot from mile post 18 to mile post 6 of lane widths in the right lane of the westbound direction.

Figure 8*e* is a heat map of speed deltas by day and longitudinal location on I-70 W. A speed delta is the difference between the average speeds of two adjacent segments. If vehicles are decelerating, the speed delta will be positive (upstream speed minus downstream speed). In this plot, only speed deltas greater than 15 mph are used so as to eliminate noise from minor changes in speed. The color scale corresponds to the duration of these speed deltas in each day. Because of the nature of the connected vehicle data segmentation, speed deltas are associated with the point connecting two adjacent segments.

In Figure 8*e*, callout "i" represents the location of the narrow lane section (also labeled in Figure 8*d*). Callout "ii" corresponds to the congestion and date shown in Figure 4*a*. In this heat map, there are recurring hot spots near the narrow lane section. On days with lane closures, such as April 19, 2017 (callout "iii"), there is a greater impact on traffic. These data show that although the narrow lane section is short, it has a recurring impact on traffic. Queues extended upstream of the work zone warning area on 4 days within this time period. When queues extend past the warning area, there is the potential for severe back-of-queue crashes to occur

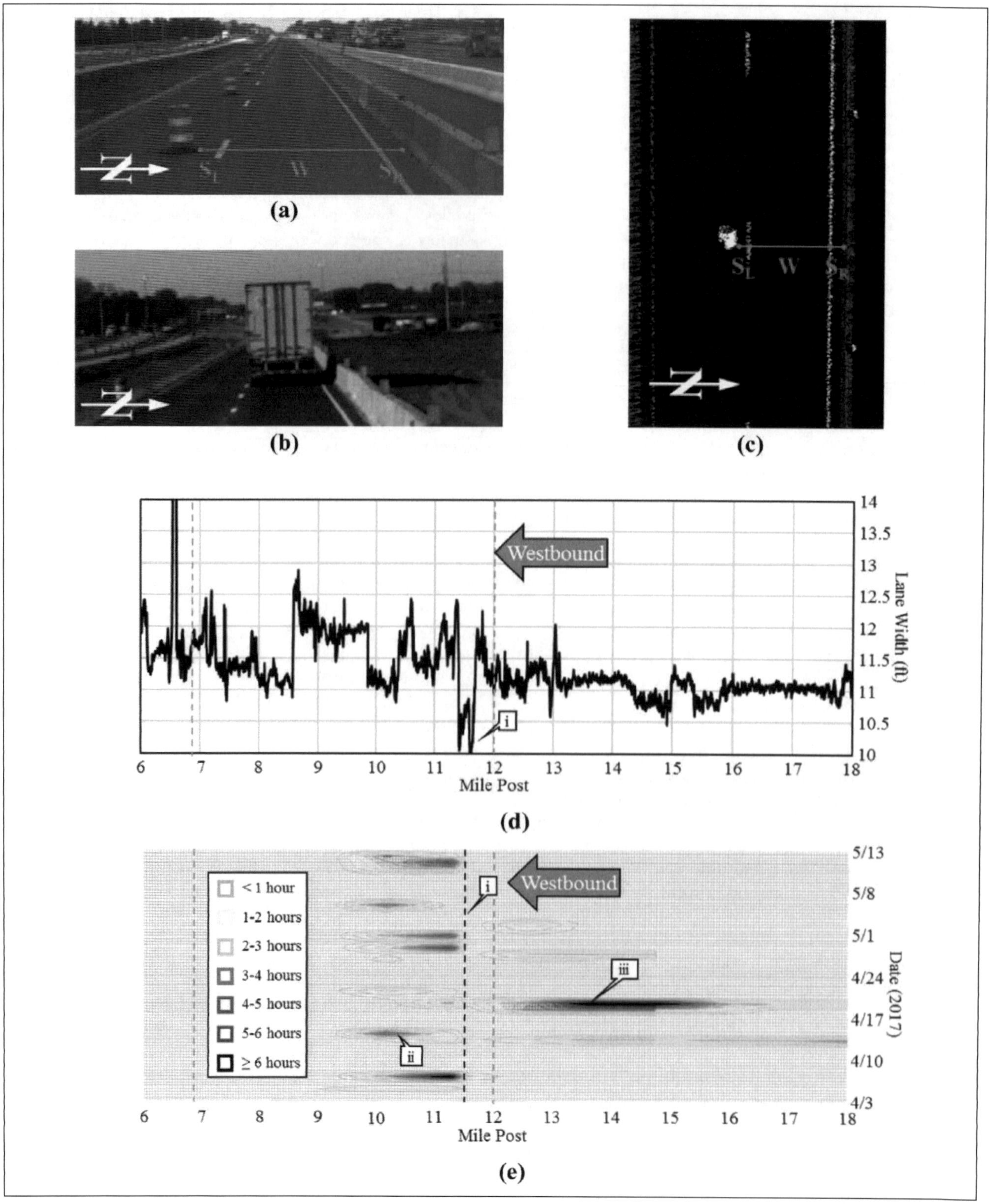

Figure 8. I-70 W lane width and traffic speed changes: (*a*) lane dimensions at I-70 W, MP 11.62; (*b*) commercial vehicle at I-70 W, MP 11.62; (*c*) point cloud at I-70 W, MP 11.62, S_L = 1.41 ft, W = 10.01 ft, S_R = 1.48 ft; (*d*) lane width by longitudinal location; (e) frequency of speed deltas ≥ 15 MPH (1-day bins).

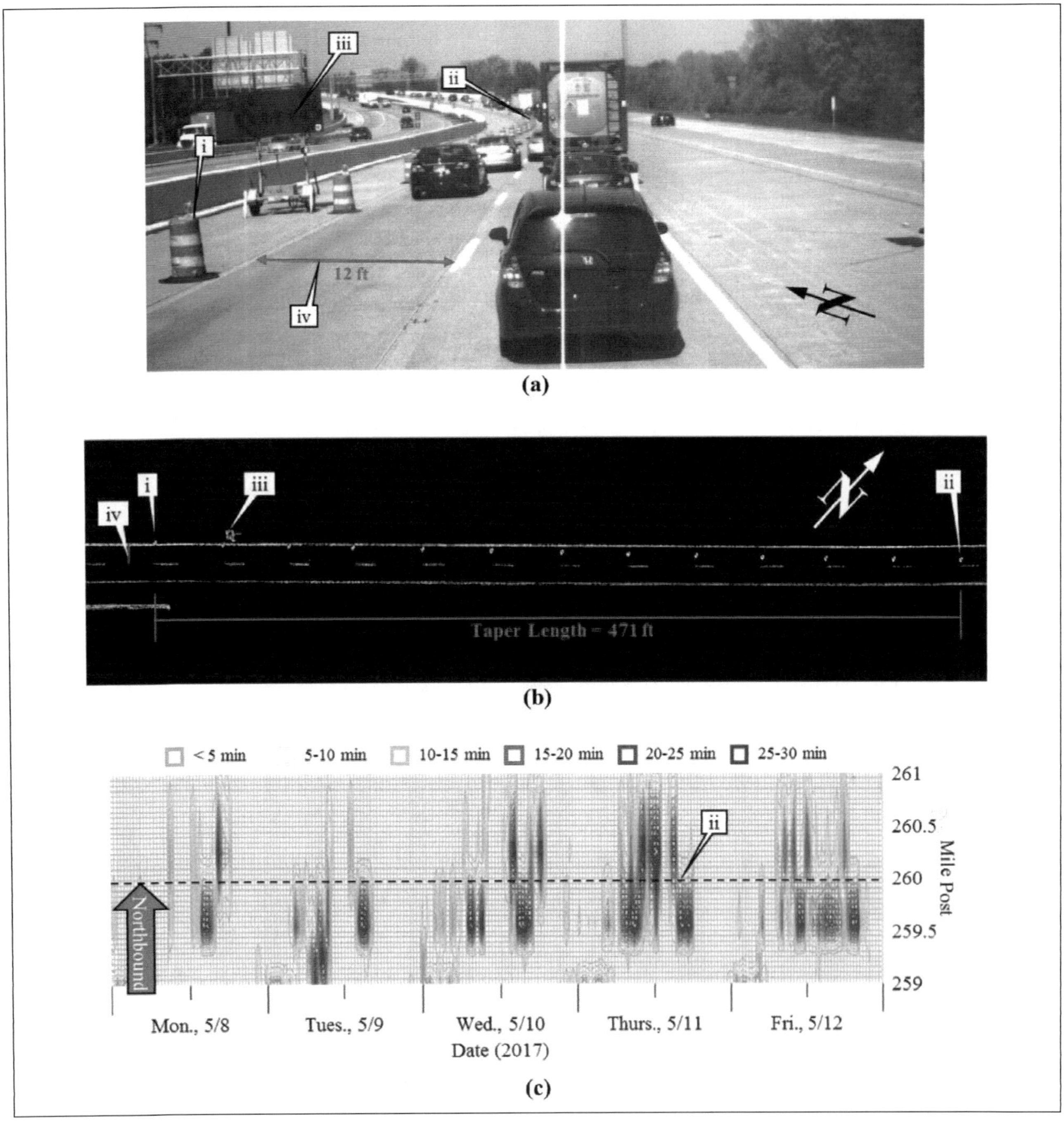

Figure 9. Merging taper on I-65 N at mile post 259.8: (*a*) view of merging taper from vehicle; (*b*) merging taper length from point cloud; (*c*) frequency of speed deltas ≥ 15 mph (30-minute bins).

because of the large speed changes and lack of warning to drivers.

It would have been time-consuming and unsafe for an inspector or contractor to manually measure the lane widths in this 12-mile section at regular intervals. It is also difficult to identify the narrow lane section visually without the presence of a wide vehicle, such as in Figure 8*b*. The data collection process took only about 20 min once the mobile LiDAR system was set up and calibrated and the work zone was reached. With the automation of data processing, the lane widths measurements can be obtained within 1 day of the data collection.

Case Study 2: Taper Length

The second case study is from a work zone on I-65 N in northern Indiana, near Chicago. This location has an AADT of 36,637, with 67% traveling northbound and 17% commercial vehicles. There is recurring congestion under non-work zone conditions. At the time of this study, traffic from I-65 N to I-80 E was detoured through this work zone because of a ramp closure at another location. At mile post 259.8 during the week of May 8, 2017, there was a lane closure and merging taper (Figure 9*a*). The posted work zone speed limit was 55 mph. According to the INDOT Work Zone Traffic Control Guidelines (*15*), the minimum required taper length is 680 ft. Alternatively, Equation 4 shows the calculation of the Manual on Uniform Traffic Control Device (*14*) recommended minimum taper length, which is 660 ft.

$$L = WS = (12 ft)(55 MPH) = 660 ft \quad (4)$$

Figure 9*b* depicts the lane markings and drums in the LiDAR data point cloud. Callouts "i" and "ii" (Figure 9*a* and *b*) correspond to the first and last drum in the merging taper, respectively. Callout "iii" corresponds to the arrow board on the left shoulder. Callout "iv" refers to the 12 ft offset of the merging taper. As measured in the LiDAR data, the actual merging taper length is only 471 ft, which is over 200 ft shorter than the INDOT minimum requirement.

Figure 9*c* is the speed delta heat map for mile post 259–261 and May 8–12 (30-minute bins). As in Case Study 1, there is a congregation of hot spots near the merging taper over the course of the week.

Conclusions

Connected vehicle speed data can be used at both the statewide and segment level to identify bottleneck locations (recurring and in work zones), localized incidents (crashes), and regional incidents (holidays and weather events). Integration of LiDAR geometric data with the connected vehicle speed data enables the diagnosis of non-conforming geometric conditions in work zones. Collection of geometric data using LiDAR can occur at highways speeds, does not require lane closures, and dramatically reduces the exposure of inspectors to traffic. This paper demonstrated that the integration of probe data and LiDAR data could be used to effectively identify unexpected congestion locations in a work zone and trace the cause of that congestion to a section of road. In the first case study, a section of road was found to have lane widths that did not meet design standards for approximately 1320 ft in a 5.2-mile construction zone (Figure 8). In addition, repeated runs (November 5, 2016, and May 2, 2017) found the lane width data was reproducible with a root mean square error of 1.6 cm (0.6 in). The second case study demonstrated the effects of a merging taper that was 200 ft shorter than the design standard. These and other work zone features were difficult to assess safely by an inspector in the field because of the high volume of traffic. In conclusion, it is recommended to use LiDAR to assess geometrics in work zones identified using connected vehicle data as having recurring congestion.

Acknowledgments

This work was supported by the Joint Transportation Research Program administered by the Indiana Department of Transportation and Purdue University. The crowd-sourced probe vehicle data was provided by INRIX. The LiDAR systems used in this research were the HDL32E and VLP16 of the Velodyne series.

References

1. INDOT. *Interstate Highways Congestion Policy*. INDOT, Indianapolis, Indiana, 2017.
2. Rouphail, N. M., Z. S. Yang, and J. Fazio. Comparative Study of Short- and Long-Term Urban Work Zones. *Transportation Research Record: Journal of the Transportation Research Board*, 1988. 1163: 4–14.
3. Andrew, L. B., and J. E. Bryden. Managing Construction Safety and Health: Experience of New York State Department of Transportation. *Transportation Research Record: Journal of the Transportation Research Board*, 1997. 1585: 9–18.
4. Bryden, J. E., and L. B. Andrew. Quality Assurance Program for Work Zone Traffic Contol. *Transportation Research Record: Journal of the Transportation Research Board*, 2001. 1745: 1–9.
5. Gambatese, J., and M. Johnson. Impact of Design and Construction on Quality, Consistency, and Safety of Traffic Control Plans. *Transportation Research Record: Journal of the Transportation Research Board*, 2014. 2458: 47–55.
6. *Highway Capacity Manual 2010*. Transportation Research Board of the National Academies, Washington, D.C., 2010.
7. *Highway Capacity Manual*. TRB, National Research Council, Washington, D.C., 2000.
8. *Special Report 209: Highway Capacity Manual*, 3rd ed. (1997 update). TRB, National Research Council, Washington, D.C., 1998.
9. Yeom, C., A. Hajbabaie, B. J. Schroeder, C. Vaughan, X. Xuan, and N. Rouphail. Innovative Work Zone Capacity Models from Nationwide Field and Archival Sources. *Transportation Research Record: Journal of the Transportation Research Board*, 2015. 2485: 51–60.
10. Weng, J., and Q. Meng. Decision Tree-Based Model for Estimation of Work Zone Capacity. *Transportation Research Record: Journal of the Transportation Research Board*, 2011. 2257: 40–50.
11. Jiang, Y. Traffic Capacity, Speed, and Queue Discharge Rate of Indiana's Four-Lane Freeway Work Zones. *Transportation Research Record: Journal of the Transportation Research Board*, 1999. 1657: 10–17.
12. Sharp, M. C., and D. W. Harwood. Effects of Taper Length on Traffic Operation in Construction Work Zones. *Transportation Research Record: Journal of the Transportation Research Board*, 1979. 703: 19–24.
13. Theiss, L., M. D. Finley, and G. L. Ullman. Merging Taper Lengths for Lane Closures of Short Duration. *Transportation Research Record: Journal of the Transportation Research Board*, 2011. 2258: 64–70.

14. Federal Highway Administration. *Manual on Uniform Traffic Control Devices*. Washington, D.C., 2009.
15. Work Zone Safety Section. Taper Length Criteria for Work Zones. *Work Zone Traffic Control Quidelines*. INDOT, Indianapolis, Indiana, 2013, p. 13.
16. El-Sheimy, N., C. Valeo, and A. Habib. *Digital Terrain Modeling: Acquisition, Manipulation, and Applications*. Artech House, Norwood, Mass., 2005.
17. Shan, J., and C. K. Toth. *Topographic Laser Ranging and Scanning: Principles and Processing*. CRC Press - Taylor & Francis Group, Boca Raton, Florida, 2009.
18. Olsen, M. J., G. V. Roe, C. Glennie, F. Persi, M. Reedy, D. Hurwitz, K. Williams, H. Tuss, A. Squellati, and M. Knodler. *NCHRP Report 748: Guidelines for the Use of Mobile LIDAR in Transportation Applications*. Transportation Research Board of the National Academies, Washington, D.C., 2013.
19. Chang, J. C., D. J. Findley, C. M. Cunningham, and M. K. Tsai. Considerations for Effective Lidar Deployment by Transportation Agencies. *Transportation Research Record: Journal of the Transportation Research Board*, 2014. 2440: 1–8.
20. Day, C. M., M. L. McNamara, H. Li, R. S. Sakhare, J. Desai, E. D. Cox, D. K. Horton, and D. M. Bullock. *2015 Indiana Mobility Report and Performance Measure Dashboards*. Purdue University, West Lafayette, Indiana, 2016.
21. Habib, A. F., A. P. Kersting, K. I. Bang, R. Zhai, and M. Al-Durgham. A Strip Adjustment Procedure to Mitigate the Impact of Inaccurate Mounting Parameters in Parallel LiDAR Strips. *The Photogrammetric Record*, Vol. 24, No. 126, 2009, pp. 171-195.
22. Habib, A., K. I. Bang, A. P. Kersting, and J. Chow. Alternative Methodologies for LiDAR System Calibration. *Remote Sensing*, Vol. 2, No. 3, 2010, pp. 874-907.
23. Lin, Y. J., R. Ravi, T. Shamseldin, M. Elbahnasawy, D. Bullock, and A. Habib. Comparative Analysis of Potential Calibration Alternatives for a Multi-Unit LiDAR System. Proceedings of the 10th International Symposium on Mobile Mapping Technology, Cairo, Egypt, 2017.
24. Yu, Y., J. Li, H. Guan, F. Jia, and C. Wang. Learning Hierarchical Features for Automated Extraction of Road Markings from 3-D Mobile LiDAR Point Clouds. *IEEE Journal of Selected Topics in Applied Earth Observations and Remote Sensing*, Vol. 8, No. 2, 2015, pp. 709–726.
25. Riviero, B., L. Diaz-Vilarino, B. Conde-Carnero, M. Soilan, and P. Arias. Automatic Segmentation and Shape-Based Classification of Reatro-Reflective Traffic Signs from Mobile LiDAR Data. *IEEE Journal of Selected Topics in Applied Earth Observations and Remote Sensing*, Vol. 9, No. 1, 2016, pp. 295–303.
26. Pu, S., M. Rutzinger, G. Vosselman, and S. O. Elberink. Recognizing Basic Structures from Mobile Laser Scanning Data for Road Inventory Studies. *ISPRS Journal of Photogrammetry and Remote Sensing*, Vol. 66, No. 6, 2011, pp. S28-S29.
27. Geiger, A., P. Lenz, and R. Urtasan. Are We Ready for Autonomous Driving? The Kitti Vision Benchmark Suite. 2012 IEEE Conference on Computer Vision and Pattern Recognition, Providence, Rhode Island, 2012, pp. 3354–3361.
28. Lato, M. J., M. S. Diederichs, D. J. Hitchinson, and R. Harrap. Harrap. Evaluating Roadside Rockmasses for Rockfall Hazards Using LiDAR Data: Optimizing Data Collecion and Processing Protocols. *Natural Hazards*, Vol. 60, No. 3, 2012, pp. 831–864.

The Standing Committee on Work Zone Traffic Control (AHB55) peer-reviewed this paper (18-00192).

The contents of this paper reflect the views of the authors, who are responsible for the facts and the accuracy of the data presented here, and do not necessarily reflect the official views or policies of the sponsoring organizations. These contents do not constitute a standard, specification, or regulation.

Article

TRR
JOURNAL OF THE TRANSPORTATION RESEARCH BOARD

Transportation Research Record
2018, Vol. 2672(16) 14–22

DOI: 10.1177/0361198118757982
journals.sagepub.com/home/trr
SAGE

Study on the Day-Based Work Zone Scheduling Problem in Urban Road Networks Based on the Day-to-Day Traffic Assignment Model

Da Yang[1,2], Xinpeng Zhao[1], Yuting Chen[1], Xi Zhang[3], and Chongshuang Chen[4]

Abstract

Many work zones exist in the urban road network and have a great negative impact on city traffic. Finding the optimal work zone schedule can minimize the negative impact of work zones on traffic. This paper focuses on the day-based work zone scheduling problem in the urban network. Existing studies on the day-based work zone scheduling problem do not consider the progression of day-to-day traffic from a non-equilibrium state to an equilibrium state during the construction period. For the first time, this paper proposes a model for day-based work zone scheduling by introducing a day-to-day traffic assignment model, in which the target of the optimization problem is minimizing the increase in travel cost caused by work zones. Numerical examples are presented to explore the variations of the optimal construction sequence for different work durations, crew numbers, and model parameter values. Some new findings are obtained in the paper. When the construction duration of each work zone is relatively short (for example, less than 20 days), the optimal scheduling will obviously change with the work duration; when all of the construction durations increase to a threshold (for example, 60 days), the optimal construction sequence will no longer change. An optimal crew number exists that can minimize the increment of travel cost caused by work zones. During the construction period, the total travel cost in the network can be decreased by guiding travelers to change their original travel habits.

In recent years, construction work has resulted in many work zones in the city road network, leading to heavy traffic congestion and increased travel costs and environmental pollution. The construction sequence for multiple projects in the urban road network has a significant impact on network travel cost (*1–3*). Therefore, it is possible to reduce the negative effect of work zones on urban traffic by finding an optimal work zone schedule.

Existing studies on work zone scheduling have mainly focused on the hour-based work zone scheduling problem, in which the duration of all the work zones in the network is less than 24 h (*4–8*). However, the duration of many work zones in the road networks is more than one day, and the scheduling of such work zones, namely, the day-based work zone scheduling problem, has not attracted much attention. So far, only a few studies have focused on the day-based work zone scheduling problem (*2,3,9,10*). Zheng et al. (*2, 3*) and Gong and Fan (*10*) calculated the traffic flow in the network using the user equilibrium (UE) traffic assignment model. In their studies, traffic volume in the network stays the same during the entire construction period. However, in real traffic, the network traffic flow will evolve day to day from a non-equilibrium state to an equilibrium state when the road capacity drops (*11*). Without consideration of the day-to-day dynamics of traffic flow, the obtained optimal work zone schedule maybe not accurate. For the first time, Lee (*9*) took into account the day-to-day dynamics of the network traffic flow; however, the traffic diversions among routes for different days were obtained through questionnaires. To better reflect the day-to-day evolution of traffic flow in construction periods, this paper adopts a day-to-day traffic assignment model to describe the daily traffic volume. The day-to-day traffic assignment method can describe the evolution process of the network traffic flow in constructions from a non-equilibrium state to an equilibrium

[1]School of Transportation and Logistics, Southwest Jiaotong University, Chengdu, China
[2]Traffic Management Research Institute of the Ministry of Public Security, Wuxi, China
[3]School of Transportation, Chongqing Jiaotong University, Chongqing, China
[4]School of Mathematics, Southwest Jiaotong University, Chengdu, China

Corresponding Author:
Address correspondence to Chongshuang Chen: ccsmars@swjtu.edu.cn

state, and has the advantages of lower cost, higher efficiency, and better accuracy.

Literature Review

Studies on work zone scheduling date back to the 1990s. The early studies (*12,13*) used a set of analytical formulas developed from queuing theory to calculate the traffic delay, and further to find the optimal schedule by minimizing the total traffic delay in the network. Fwa et al. (*12*) proposed an optimization method based on a genetic algorithm (GA) to schedule ten work zones with three work crews in one day, which was subsequently expanded to schedule one hundred work zones with six work crews by Muntasir (*13*). However, these early studies assumed road sections were isolated and ignored the traffic interactions among adjacent links.

After 2000, researchers started to investigate the work zone scheduling problem considering the traffic distribution among routes in the road network. At this stage, studies on work zone scheduling mainly focused on the case that the work zones only exist for several hours, and all the construction jobs need to be finished within 24 h (*4,5,7,8,12,14–16*). Among these studies, some researchers (*4,5,7,8*) directly adopted microscopic traffic simulation tools, such as PARAMICS and VISSIM, to describe the traffic volume and cost in the network. Chang et al. (*1*) adopted both the dynamic traffic assignment method and all-or-nothing assignment method to calculate the peak-hour traffic volume and traffic delay for different work zone schedules and then obtain the optimal schedule. The above studies have a strong assumption that all the construction projects have the same duration, which is not consistent with practice. Tang and Chien (*14,16*) adopted the UE traffic assignment method to determine the time-varying traffic volume and further achieved an optimal work zone schedule taking into account the work zone configuration, although the study was conducted on a small-scale rural road network.

However, work zones lasting for several days, or even more than one hundred days, exist widely in city road networks, and finding the optimal work zone schedule for these work zones, called the day-based work zone scheduling problem in this paper, has not attracted much attention. So far, only a few studies have investigated the day-based work zone scheduling problem (*2,3,9,10*). Zheng et al. (*2,3*) calculated the traffic delay based on a UE traffic assignment model using a month as the unit. However, using the month as the unit to assign traffic has some deficiencies. First, this method only can deal with the problem in which the duration of work zones were integer times of a month, although the duration of many work zones do not meet this requirement in practice. Second, this method assumed that the traffic flow was in an equilibrium state during the whole month, which ignored the day-to-day dynamics of traffic when work zones exist. For the first time, Lee (*9*) considered the day-to-day dynamics of the traffic in the network. However, in this study, the turning percentage among routes was obtained through questionnaires. The questionnaire method has some deficiencies. First, many samples need to be collected, which creates a huge workload and takes time. Second, it is difficult to extend the questionnaire method to a large-scale network, because it cannot accurately describe the interaction among multiple work zones. Recently, Gong et al. (*10*) investigated the optimal schedule for work zones in the network; however, the daily traffic volume was still calculated using the UE distribution method, without considering the day-to-day dynamic characteristics of the network traffic flow.

In summary, existing studies on work zone scheduling cannot capture the day-to-day dynamic evolution of the network traffic flow during the construction period. Therefore, to analyze the daily traffic delay and obtain the optimal work zone schedule more accurately, a new method needs to be applied.

Day-to-Day Based Work Zone Scheduling Model

Work Zone Scheduling Model

Let $G(N, M)$ denote a fully connected directed traffic network, consisting of N nodes and M links. W denotes the number of origin-destination (OD) pairs in the network, d denotes the fixed demand between the OD pair r and s, U denotes the number of work zones, and C denotes the number of crews.

When considering the day-to-day traffic evolution in the network, the impact of work zones on travel cost can be divided into two phases. The first phase is the construction period, and the second phase is from the completion of all the construction jobs to the time that the traffic in the network reaches an equilibrium state. In the first phase, the existence of work zones reduces road capacities, leading to increased travel cost. In this phase, the increased travel cost is directly produced by work zones; this increased cost is called the direct cost. In the second phase, although the work zones are removed from the network, the traffic still needs a period of time to reach an equilibrium state, and the travel cost is still in variation. In this phase, the cost change is not produced directly by work zones, and is called the indirect cost. In existing studies on the work zone scheduling problem the indirect cost was ignored, which may produce an inaccurate optimal work zone schedule.

The work zone scheduling problem can be summarized as a non-linear program problem (*1,8*). This paper takes into account both the direct and indirect costs and adopts the cost increment caused by work zones as the objective function. The proposed work zone scheduling model is as follows:

$$min\ P(j) = \sum_{i=1}^{n} \sum_{t_i=1}^{T_i} \boldsymbol{c}\left(\boldsymbol{x}^{t_i}\right)' \cdot \boldsymbol{x}^{t_i} - \left(\sum_{i=1}^{n} T_i + L \right) \cdot \boldsymbol{c}\left(\boldsymbol{x}_0\right)' \cdot \boldsymbol{x}_0 \qquad (1)$$

S.T.

$$\sum_{k} f^{r,s}_{t_i,k} = d_{r,s} \ t_i \in [1,T_i], \forall r,s,k \quad (2)$$

$$\sum_{k} f^{r,s}_{k} = d_{r,s} \ \forall r,s,k \quad (3)$$

$$x^{t_i} = \sum_{r}\sum_{s}\sum_{k} f^{r,s}_{t_i,k}\delta^{r,s}_{a,k} \ \forall r,s,t_i,k,a \quad (4)$$

$$x_0 = \sum_{r}\sum_{s}\sum_{k} f^{r,s}_{k}\delta^{r,s}_{a,k} \ \forall r,s,k,a \quad (5)$$

$$f^{r,s}_{t_i,k} \geq 0 \ t_i \in [1,T_i], \forall r,s,k \quad (6)$$

$$f^{r,s}_{k} \geq 0 \ \forall r,s,k \quad (7)$$

where, $P(j)$ denotes the increased travel cost caused by work zones for the schedule j, i denotes the ith construction phase, t_i denotes the tth day in the construction phase i, T_i denotes the number of days in the construction phase i, L denotes the number of days in the second phase, x^{t_i} denotes the link flow vector in the tth day for the construction phase i, x_0 denotes the link flow vector before construction works start, $c(\cdot)$ denotes the travel cost vector calculated using the Bureau of Public Roads (BPR) function, $d_{r,s}$ denotes the fixed demand between the OD pair (r, s), $f^{r,s}_{t_i,k}$ denotes the flow of the route k between the OD pair (r, s) in the tth day of the construction phase i, $f^{r,s}_{k}$ denotes the flow of the route k between the OD pair (r, s) before construction works start, and $\delta^{r,s}_{a,k}$ is 1 if the link a belongs to the route k between the OD pair (r, s) and is 0 otherwise.

In the model, a construction phase refers to a capacity state in the network (*10*). In different construction phases, the highway capacity in the network has different values. A new construction task starting or ending in the network will produce a new construction phase. For example, assume that there are two construction jobs which need to be finished by two crews; the first job needs 30 construction days and the second job needs 40 construction days, and the two construction jobs start simultaneously. There will be three construction phases for this case. Construction phase one is from the starting of the two construction jobs to the end of the first job, so it lasts 30 days. Construction phase two is from the end of the first job to the end of the second job, so it lasts 10 days. Construction phase three is from the end of the second phase to the time the traffic flow reaches an equilibrium state.

In the model, Equation 1 is the objective function, which consists of two parts. The first part is the sum of the direct cost and indirect cost caused by work zones, and the second part is the total travel cost in all construction phases if no work zones are located in the network. The difference between the two parts reflects the total increment of travel cost caused by work zones. Equations 2–7 are the constraints, which indicate that the traffic demand should be non-negative, and the total traffic demand for the OD pair (r, s) equals the total demand on all the paths from r to s. The flow in the model meets the day-to-day traffic evolution in the following context.

Solving the work zone scheduling problem is difficult and time-consuming (*1*); so far, the methods used to solve this problem are heuristic algorithms, including Tabu search algorithm (*1*), simulated annealing algorithm (*6*), ant colony algorithm (*9*) and GA (*4,5,7,8,12–17*).

This paper adopts the GA to solve the proposed model. Compared with other algorithms, it has the advantages of global search ability, small possibility of trapping in local optimal, fast convergence speed, discrete variable feature, and so forth. The classical GA was modified to be applicable to the proposed model. First, in the genetic representation, a number was assigned for each work zone and take "0" as the task dividing line for each crew. Second, in determining the initial population, several individuals were randomly generated as the start of the next iteration. Since it is a minimization problem, the individual with the higher fitness has the greater possibility of passing genes to the next generation. Thus, the reciprocal of the objective function was taken as the fitness function, namely, $1/P(j)$. Third, in genetic rules, individuals are selected in the next generation by fitness ratio, which makes the individuals with higher fitness have the higher inheriting possibility. To reduce the possibility of converging to a local optimal solution, partially mapped crossover was used, reversing mutation and other rules with a proper probability to update individuals and increase the diversity of solutions.

Day-to-Day Traffic Assignment Model

The day-to-day traffic assignment model focuses on the daily evolution process of traffic in the road network. Compared with the classical UE model, it can describe the day-to-day traffic development from a non-equilibrium state to an equilibrium state. To illustrate the differences among them, Figure 1, a and b, respectively, display the link flow variations for the UE models and day-to-day models. It can be seen that, when lanes are closed because of work zones or opened after work zones are removed, the link flow calculated using the day-to-day model fluctuates at first, and gradually evolves from a non-equilibrium state to an equilibrium state. In contrast, the link flow calculated using the UE model reaches an equilibrium state on the first day of the construction and does not have the day-to-day evolution characteristic.

The existing day-to-day traffic assignment method can be classified into two categories: the path-based model (*18–22*) and the link-based model (*11,23,24*). Compared with the path-based model, the link-based model can describe the flow variations on links. In calculating the flow rate in the network, the initial inputs of this type of

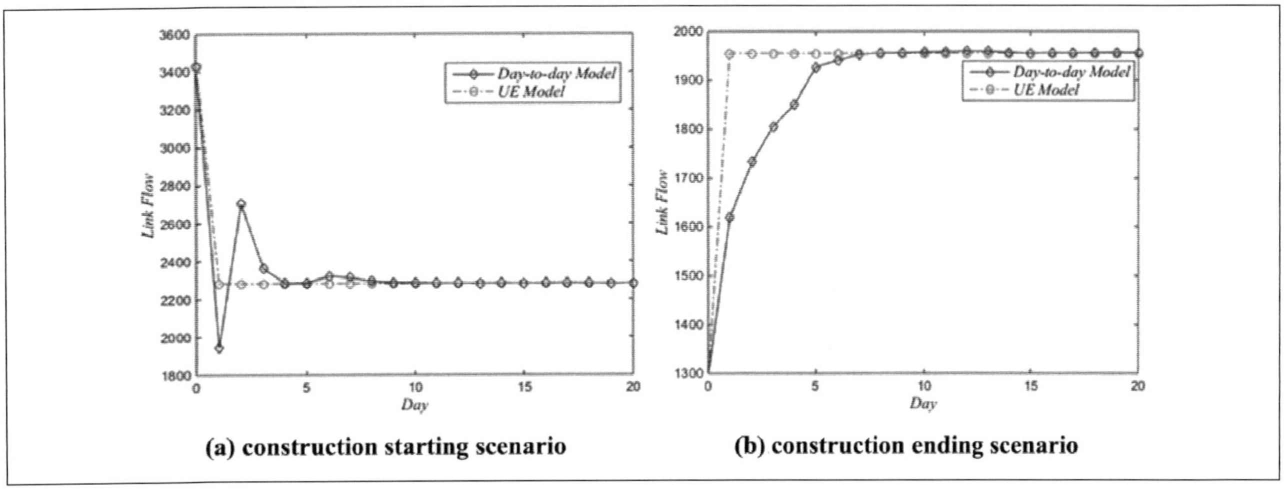

Figure 1. Day-to-day traffic evolvement during road construction.

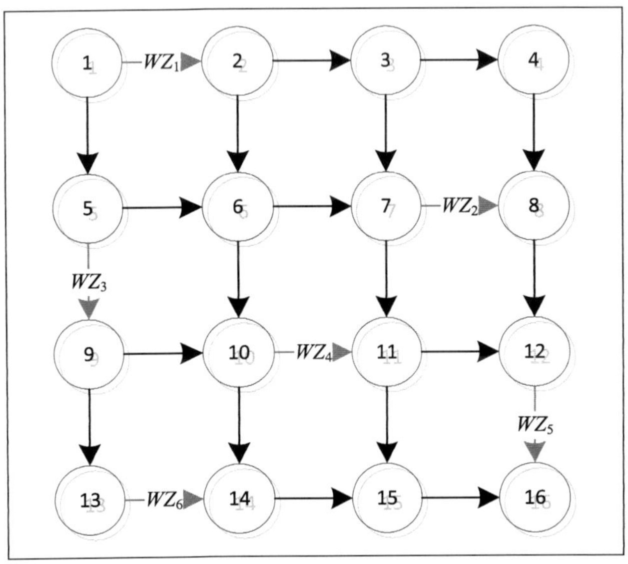

Figure 2. Test network.

model are link flow values, which are in practice much easier to obtain than the path flow. Moreover, the path interdependence issue can be avoided in the link-based model (*11*). Therefore, this paper adopts the day-to-day link-based model proposed by He et al. (*11*) to calculate the traffic flow in the network when work zones exist, and the formula of the model is as follows:

$$x^{t+1} = x^t + \alpha \cdot \left(y^{t+1} - x^t \right) \tag{8}$$

where, α ($0 < \alpha \le 1$) denotes the step-size, x^t denotes the link flow vector at the day t, x^{t+1} denotes the link flow vector at the day t+1, and y^{t+1} denotes the "target" flow pattern at the day t+1 that can be solved by the following equation:

$$\min \lambda c\left(x^t\right)' y^{t+1} + (1-\lambda) D\left(x^t, y^{t+1}\right) \tag{9}$$

where, λ denotes a positive scalar ($0 < \lambda < 1$), $c(x^t)$ denotes a link cost vector, and $D(x^t, y^{t+1})$ denotes a measure of the distance between the target flow y^{t+1} and the current flow x^t.

Use the following equation of $D(x^t, y^{t+1})$ to compute the distance,

$$D\left(x^t, y^{t+1}\right) = \sum_{a \in L} \int_{x_a^t}^{y_a^{t+1}} \left(c_a(w) - c_a\left(x_a^t\right)\right) dw \tag{10}$$

According to He et al. (*11*), the first fraction in Equation 9 is a linear programming problem, namely, minimizing the total network cost under a given fixed link cost vector $c(x^t)$. The target of the problem is to find the shortest path for each OD pair and assign all traffic demands to the shortest path adopting the all-or-nothing assignment method. In finding the shortest path, travelers are reluctant to change their original paths because of habits, personal preferences, and so forth, so the second fraction in Equation 9 is used to represent the travelers' inertia in changing from the original paths to the new shortest path. In our model, it needs to be explained that if the construction phase of the adjacent two days is not changed, then $c(x^t)$ denotes the t-th day actual link cost vector. If the construction phase of the adjacent two days is changed, then $c(x^t)$ denotes the link cost vector which can be calculated by the t-th day link in the background of the next construction phase.

Numerical Examples

Test Network

The test network used in this paper is displayed in Figure 2. It is a fully connected directed graph, including 1 OD pair

Table 1. Link Traffic Information

Link	Free travel time	Capacity without work zone	Capacity with work zone
1→2	1300	3600	1320
2→3	1300	3600	3600
3→4	1300	3600	3600
1→5	1300	3600	3600
2→6	1300	1200	1200
3→7	1350	1200	1200
4→8	1400	3600	3600
5→6	1400	2400	2400
6→7	1400	2400	2400
7→8	1400	2400	1480
5→9	1300	3600	2520
6→10	1300	1200	1200
7→11	1350	1200	1200
8→12	1300	3600	3600
9→10	1300	1200	1200
10→11	1350	2400	1480
11→12	1350	1200	1200
9→13	1400	3600	3600
10→14	1350	1200	1200
11→15	1400	1200	1200
12→16	1400	3600	1320
13→14	1300	3600	1320
14→15	1400	3600	3600
15→16	1300	3600	3600

Table 2. Work Duration Setup

Scenario	WZ_1	WZ_2	WZ_3	WZ_4	WZ_5	WZ_6	R (%)	D (%)
1	10	12	13	15	18	20	35.02	16.52
2	20	24	26	30	36	40	32.78	11.26
3	40	48	52	60	72	80	16.43	5.64
4	50	60	65	75	90	100	14.83	4.06
5	60	72	78	90	108	120	13.41	3.86
6	80	96	104	120	144	160	10.06	3.57

(from node 1 to node 16), 16 nodes and 24 links. Meanwhile, it is assumed that the six work zones are randomly located on the six links in the network. The traffic demand is set to a fixed value of 6500; for the capacity and travel time information see Table 1.

Work Duration

In this subsection, the impact of work durations on the optimal schedule is analyzed. Assume that six work zones are assigned to three crews, and the duration of each work zone is shown in Table 2. The six scenarios in which the work duration increases gradually are designed to investigate how the travel cost and optimal schedule change with variations of the work duration.

Figure 3 exhibits the day-to-day development of the total cost and the Gantt chart of the optimal construction schedule for the six scenarios. From Figure 3, it can be seen that the starting and ending of the work zone can cause travel cost fluctuations in the network. When the work duration is relatively short (e.g., Scenario 1), the cost fluctuation caused may be maintained for the entire construction period and superimpose with the new fluctuation wave caused by the next construction work. In comparison, when the construction period is relatively long (e.g., Scenario 3), the cost fluctuation wave in a construction phase will vanish and the traffic can reach an equilibrium state during the construction phase. Therefore, the travel cost will change for different work durations and further result in the change of the optimal work zone schedule. The six work zone scenarios produce four optimal work zone schedules, in which Scenario 1, Scenarios 2 and 3, Scenarios 4, and Scenarios 5 and 6, respectively, correspond to an optimal schedule. Work durations were further increased based on Scenario 6, but it was found that the optimal work zone schedule will not change further; this reveals that even though the optimal work zone schedule will change with the variations of work durations, a stable threshold exists. The reason for this phenomenon is that the equilibrium state is a stable stage, and increasing the time of the equilibrium state will not change the optimal schedule.

To further illustrate the relationship between work duration and the optimal work zone schedule, the ratio R of the total non-equilibrium cost to the total cost in the construction period is defined as follows:

$$R = \frac{\sum_{i=1}^{n-1}\sum_{s_i=1}^{S_i} c\left(x^{s_i}\right)' \cdot x^{s_i}}{\sum_{i=1}^{n-1}\sum_{t_i=1}^{T_i} c\left(x^{t_i}\right)' \cdot x^{t_i}} \quad (11)$$

where, s_i denotes the s_ith non-equilibrium day in the construction phase i, and S_i denotes the total number of non-equilibrium days in the construction phase i.

The bigger value of R indicates the higher proportion of the travel cost fluctuation caused by work zones in the total cost, which further represents the importance of considering the day-to-day dynamic characteristics of the traffic flow in the work zone scheduling problem. Table 2 also presents the R-value for the six scenarios. For the scenarios with short work durations, the value of R is significantly higher, such as Scenarios 1 and 2, which proves the necessity of considering the day-to-day dynamic characteristics of traffic flow for the scenarios with short work durations. When the duration of work zones is more than 80 days, the values of R are less than 10%, which means the impact of the day-to-day characteristic of traffic flow on optimal work zone schedule has become very small. In addition, the R-value can also be used to evaluate the relationship between the optimal work zone schedule and work durations. After R <14% (Scenario 5), the increment of work durations will no longer change the optimal work zone schedule.

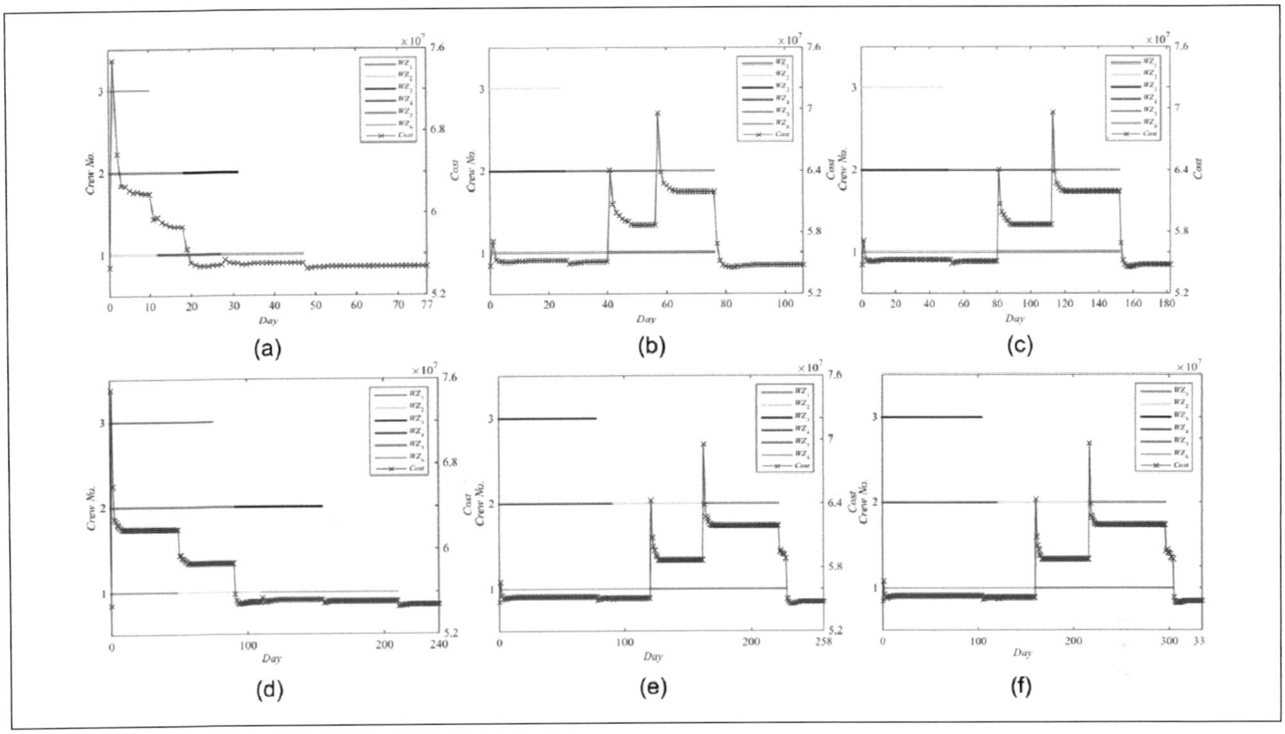

Figure 3. Total travel cost variation and Gantt chart for the optimal work zone schedule.

To illustrate the difference between the day-to-day and UE traffic assignment models, D is used, that is the difference of the increased costs calculated using the day-to-day model and UE model respectively, to investigate the application range of the two kinds of models. The equation of D can be expressed as follows:

$$D = \frac{P - Q}{Q} \tag{12}$$

where P denotes the increased cost calculated by the day-to-day traffic assignment model for a given schedule j, and Q denotes the cost calculated by the UE traffic assignment model for the same schedule.

The values of D are shown in Table 2. The smaller value of D means the closer cost results obtained by the two models. In Scenario 1, the results calculated by the two methods have 16.52% difference, while in Scenario 6, the value of D is only 3.57%. This result indicates that when the construction period is long, the cost difference calculated using the two models is small. For long work durations, the UE model should be used instead of the day-to-day model to calculate the travel cost caused by work zones, since the UE model uses less computation time and resources.

Number of Crews

Based on Scenario 1 in Table 2, this subsection discusses the variations of the optimal schedule for the crew number changing from 1 to 6, as shown in Figure 4. From Figure 4, it can be seen that when the crew number is small (e.g., the crew number is 1 and 2), the total travel cost does not increase much because the number of construction sites in the network is few. In comparison, when the crew number is more than 3, the total cost increases sharply. In such a case, congestion is heavy during the construction period, but the total construction time is much shorter.

In addition, calculated are the increased total cost P, the total construction duration, and the R-value, respectively, for the various crew numbers, as shown in Figure 5. Figure 5a displays the variations of the increased total cost P and R-value with the change of crew number. It indicates that the crew number does not has a linear relationship with the total travel cost and an optimal crew number that can minimize the increased travel cost caused by work zones exists. In the given examples, when the crew number is 2, the increased total cost caused by work zones is minimal. For the value of R, when the crew number is less than 4, the R-value is in the range of 30–35%, and when the crew number is up to 6, the R-value increases to 71%; this indicates that the traffic fluctuates greatly and travelers need to adjust their path choices frequently when the crew number is big. This phenomenon happens because too many construction works start at the same time in the network, which should be avoided in practice.

In Figure 5b, for the total work duration, the crew number also has a non-linear relationship with the total work duration. As observed from Figure 5b, when the crew number increases

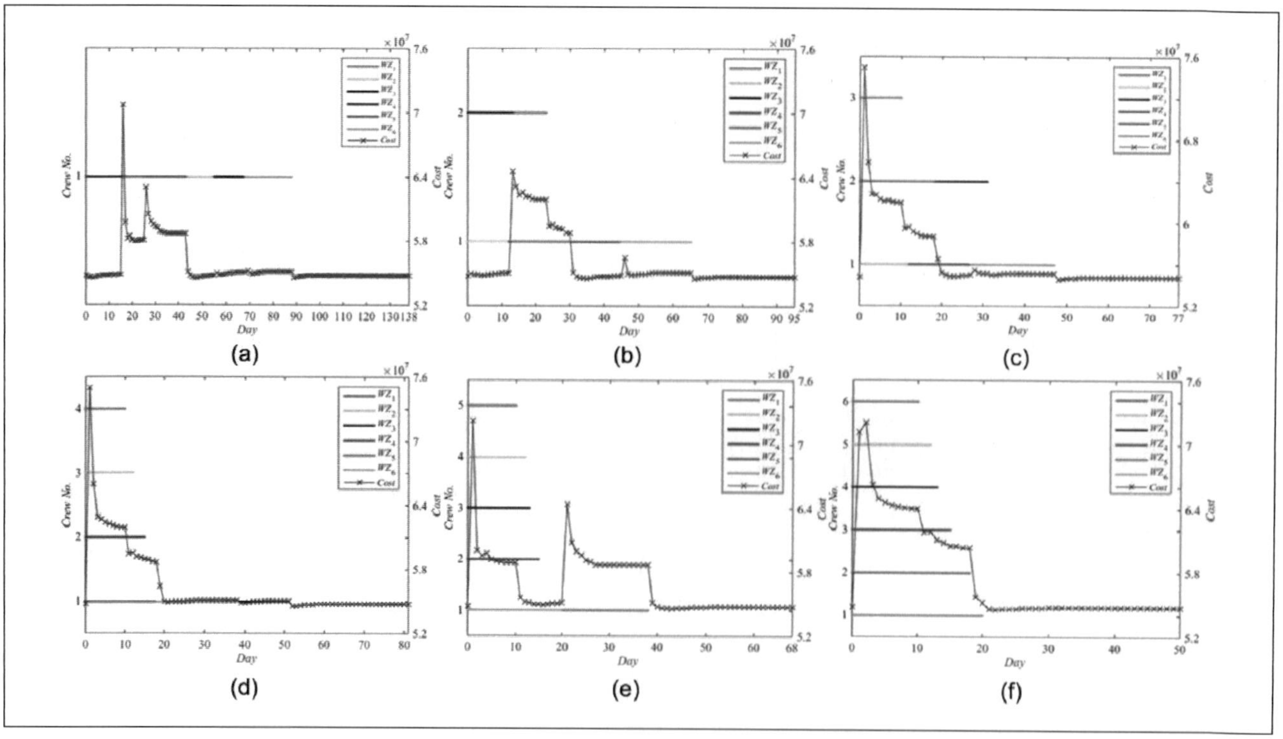

Figure 4. Total travel cost variation and Gantt chart for the optimal work zone schedule under different crew numbers.

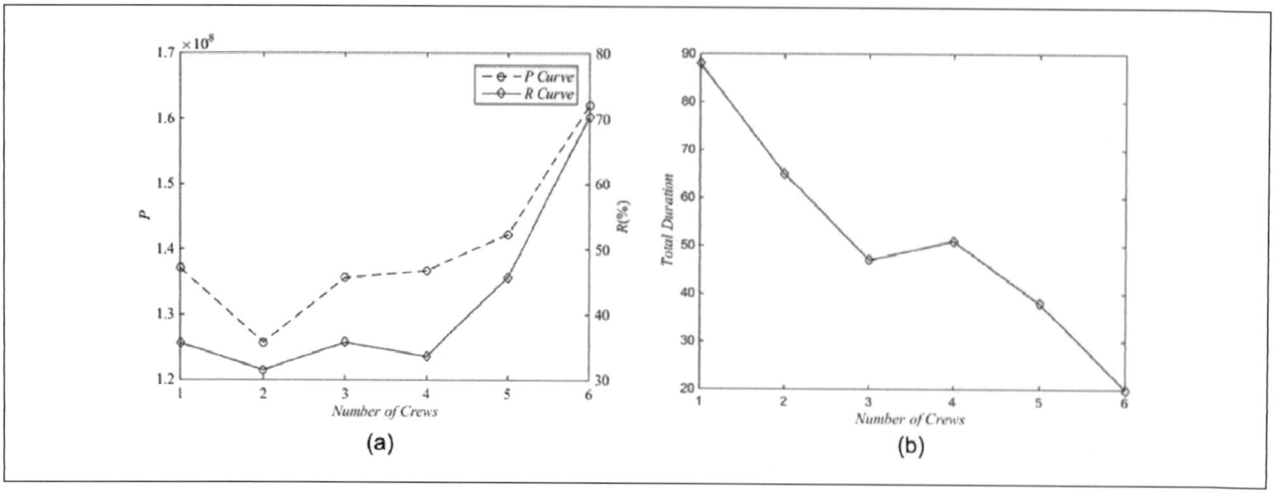

Figure 5. Impact of the crew number: (a) the increased total cost (P) and the ratio (R); (b) R values.

from 3 to 4, the total work duration does not reduce but increases 4 days.

Parameter Analysis

The parameter α in the used day-to-day traffic assignment model (*11*) represents travelers' inertia of changing from the current path to a new path. This subsection will explore how the parameter affects the travel cost and optimal work zone schedule.

Let $\lambda = 0.7$ as suggested by He et al. (*11*), increase the value of α from 0.3 to 0.9 at the interval of 0.2 gradually. Figure 6 exhibits the total increased cost and the number of the non-equilibrium days for different α values. As shown in Figure 6a, when $\alpha = 0.5$, the total increased cost of the network is smallest. This result reveals that the negative impact of work zones on traffic can be reduced by guiding the choices of travelers to make α close to 0.5 in the road construction, such as pushing more route suggestion information.

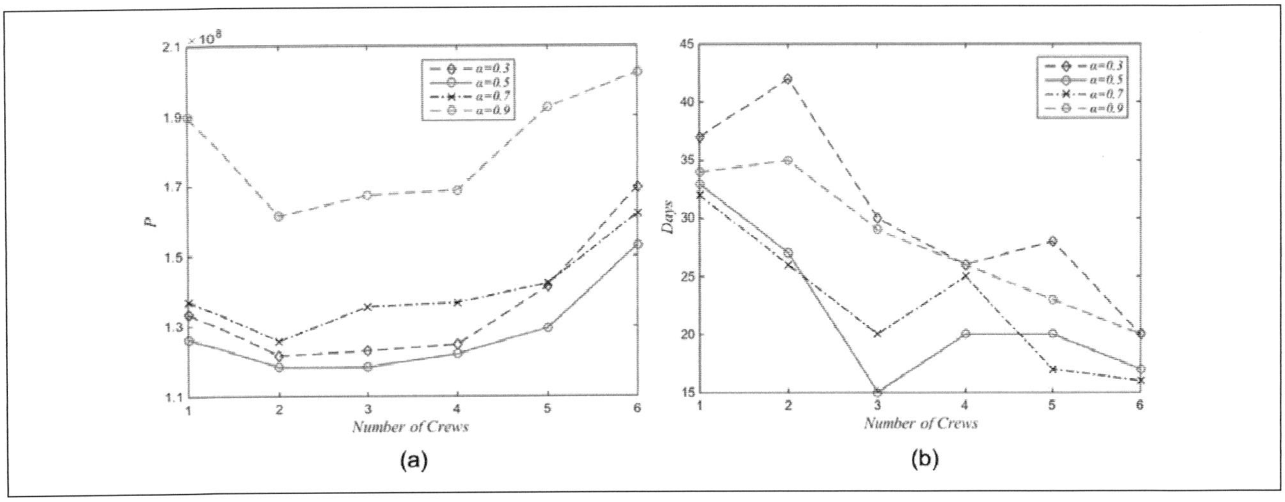

Figure 6. Impact of the model parameter α: (a) increased total travel cost; (b) number of non-equilibrium days.

Figure 6b shows that when $\alpha = 0.3$ and 0.9, the number of the non-equilibrium days are more than $\alpha = 0.5$ and 0.7. As mentioned above, the larger α means travelers tend to change their current paths, resulting in more severe network flow fluctuation, and further increase the non-equilibrium days, so it is easily understood that the non-equilibrium flow fluctuation is big for $\alpha = 0.9$. In the case of $\alpha = 0.3$, although link flow does not fluctuate intensively as the cases $\alpha = 0.5$, 0.7 and 0.9, the number of the non-equilibrium days still increases because of the slow convergence speed of traffic flow and the increased total work duration. This phenomenon implies that α should not be too small or too big to reduce the traffic fluctuation.

Conclusions and Future Work

It is common that more than one construction project exists in the urban road network simultaneously, which raises the problem of work zone scheduling. By finding the optimal work zone schedule, the negative effect of work zones on city traffic can be reduced. This paper focuses on the day-based work zone scheduling problem that has not been well studied, by introducing a day-to-day traffic assignment method. The following conclusions are drawn in the paper:

1) Work durations have a significant impact on the optimal work zone schedule. However, a duration threshold (e.g., 60 days in the numerical examples) exists, and after this threshold increasing the work durations will no longer change the optimal work zone schedule.
2) When the work duration is more than 60 days, the travel cost calculated using the day-to-day model and UE model has a negligible difference (less than 4%). Therefore, the UE model should be used to replace the day-to-day model to obtain the travel cost when the work duration is more than 60 days, in studying the day-based work zone schedule problem.
3) An optimal crew number with the optimal work zone schedule can minimize the total increased travel cost caused by work zones in the network. More crews can shorten the total work duration but do not mean less total travel cost.
4) When the model parameter α is close to 0.5, the increased total cost caused by work zones in the network is the smallest. To reduce the negative impact of work zones on traffic, the model parameter α should be close to 0.5; this can be done by guiding travelers to change their travel path choices, for example pushing more route suggestion information.

There are also some limitations in this paper. First, this paper is a theoretical study and does not include any attempt at empirical validation of the model. Further, regarding practical choices and limitations, the agency cost is not taken into account and the day-to-day model assumes that the travelers are rational in the decision-making process, while in practice travelers have bounded rationality. In the future, we will improve the study by considering the agency cost, the travelers' bounded rationality, and by examining the model using practical examples.

Acknowledgments

This work was supported by National Science Foundation of China (No. 51408509) and Fundamental Research Funds for the Central Universities (Grant No. 2682016CX046).

References

1. Chang, Y. Y., O. B. Sawaya, and A. K. Ziliaskopoulos. A Tabu Search Based Approach for Work Zone Scheduling. Presented at 80th Annual Meeting of the Transportation Research Board, Washington, D.C., 2001.
2. Zheng, H., C. Cai, E. J. Nava, V. V. Dixit, Y.-C. Chiu, E. Radwan, and D. Ismart. Optimization of Renewal-Based Project Scheduling in an Urban Network. Presented at 91st Annual Meeting of the Transportation Research Board, Washington, D.C., 2012.
3. Zheng, H., E. Nava, and Y. C. Chiu. Measuring Networkwide Traffic Delay in Schedule Optimization for Work-Zone Planning in Urban Networks. *IEEE Transactions on Intelligent Transportation Systems*, Vol. 15, No. 6, 2014, pp. 2595–2604.
4. Cheu, R., and W. Ma. An Improved Genetic Algorithm-Simulation Methodology for Lane Closure Scheduling. Presented at 81st Annual Meeting of the Transportation Research Board, Washington, D.C., 2001.
5. Cheu, R., Y. Wang, and T. Fwa. Genetic Algorithm-Simulation Methodology for Pavement Maintenance Scheduling. *Computer-Aided Civil and Infrastructure Engineering*, Vol. 19, No. 6, 2004, pp. 446–455.
6. Chien, S., Y. Tang, and P. Schonfeld. Optimizing Work Zones for Two-Lane Highway Maintenance Projects. *Journal of Transportation Engineering*, Vol. 128, No. 2, 2002, pp. 145–155.
7. Ma, W., R. Long Cheu, and D. H. Lee. Scheduling of Lane Closures Using Genetic Algorithms with Traffic Assignments and Distributed Simulations. *Journal of Transportation Engineering*, Vol. 130, No. 3, 2004, pp. 322–329.
8. Wang, Y., R. L. Cheu, and T. F. Fwa. Highway Maintenance Scheduling Using Genetic Algorithm with Microscopic Traffic Simulation. Presented at 81st Annual Meeting of the Transportation Research Board, Washington, D.C., 2002.
9. Lee, H. Y. Optimizing Schedule for Improving the Traffic Impact of Work Zone on Roads. *Automation in Construction*, Vol. 18, No. 8, 2009, pp. 1034–1044.
10. Gong, L., and W. Fan. Optimizing Scheduling of Long-Term Highway Work Zone Projects. *International Journal of Transportation Science and Technology*, Vol. 5, No. 1, 2016, pp. 17–27.
11. He, X., X. Guo, and H. X. Liu. A Link-Based Day-To-Day Traffic Assignment Model. *Transportation Research Part B: Methodological*, Vol. 44, No. 4, 2010, pp. 597–608.
12. Fwa, T., R. Cheu, and A. Muntasir. Scheduling of Pavement Maintenance to Minimize Traffic Delays. *Transportation Research Record: Journal of the Transportation Research Board*, 1998. 1650: 28–35.
13. Muntasir, A. *Scheduling of Pavement Maintenance Activities*. MS thesis. Department of Civil Engineering, National University of Singapore, 1998.
14. Chien, S. I. J., and Y. Tang. Scheduling Highway Work Zones with Genetic Algorithm Considering the Impact of Traffic Diversion. *Journal of Advanced Transportation*, Vol. 48, No. 4, 2014, pp. 287–303.
15. Fwa, T., W. Chan, and C. Tan. Genetic-Algorithm Programming of Road Maintenance and Rehabilitation. *Journal of Transportation Engineering*, Vol. 122, No. 3, 1996, pp. 246–253.
16. Tang, Y., and S. Chien. Optimization of Work Zone Schedule Considering Time-Varying Traffic Diversion. Presented at 89th Annual Meeting of the Transportation Research Board, Washington, D.C., 2009.
17. Chan, W., T. Fwa, and C. Tan. Road-Maintenance Planning Using Genetic Algorithms. I: Formulation. *Journal of Transportation Engineering*, Vol. 120, No. 5, 1994, pp. 693–709.
18. Smith, M. J. The Stability of a Dynamic Model of Traffic Assignment-An Application of a Method of Lyapunov. *Transportation Science*, Vol. 18, No. 3, 1984, pp. 245–252.
19. Nagurney, A., and D. Zhang. Projected Dynamical Systems in the Formulation, Stability Analysis, and Computation of Fixed-Demand Traffic Network Equilibria. *Transportation Science*, Vol. 31, No. 2, 1997, pp. 147–158.
20. Friesz, T. L., D. Bernstein, N. J. Mehta, R. L. Tobin, and S. Ganjalizadeh. Day-To-Day Dynamic Network Disequilibria and Idealized Traveler Information Systems. *Operations Research*, Vol. 42, No. 6, 1994, pp. 1120–1136.
21. Bertsekas, D. P., and E. M. Gafni. Projection Methods for Variational Inequalities with Application to the Traffic Assignment Problem. *Nondifferential and Variational Techniques in Optimization*, Vol. 27, 1982, pp. 139–159.
22. Zhang, D., and A. Nagurney. On the Local and Global Stability of a Travel Route Choice Adjustment Process. *Transportation Research Part B: Methodological*, Vol. 30, No. 4, 1996, pp. 245–262.
23. Guo, X., and H. X. Liu. Bounded Rationality and Irreversible Network Change. *Transportation Research Part B: Methodological*, Vol. 45, No. 10, 2011, pp. 1606–1618.
24. He, X., and H. X. Liu. Modeling the Day-To-Day Traffic Evolution Process After an Unexpected Network Disruption. *Transportation Research Part B: Methodological*, Vol. 46, No. 1, 2012, pp. 50–71.

The Standing Committee on Work Zone Traffic Control (AHB55) peer-reviewed this paper (18-02584).

Article

TRR
JOURNAL OF THE TRANSPORTATION RESEARCH BOARD

Transportation Research Record
2018, Vol. 2672(16) 23–34

DOI: 10.1177/0361198118758056
journals.sagepub.com/home/trr
SAGE

Traffic Flow Modeling of Diverse Work Zone Activities

Nipjyoti Bharadwaj[1], Praveen Edara[1], Carlos Sun[1], Henry Brown[1], and Yohan Chang[1]

Abstract

What effect does work activity type have on traffic conditions in a work zone? This question has still not been answered satisfactorily in practice. Without knowing the true effect a work activity has on traffic, practitioners are forced to make assumptions while scheduling work. This paper attempts to answer this question by studying the traffic flow characteristics, that is, traffic speed versus flow curves, capacity reduction factors, and free-flow speed reduction factors, for various activities related to construction and maintenance. The importance of the speed–flow curves and reduction factors for work zone planning is also stressed in the latest edition of the Transportation Research Board's *Highway Capacity Manual*. This manual recommends capacity and speed reduction factors for work zones, yet does not include specific guidance for including the impact of work activities. Thre e traffic stream models, Gipps, Newell–Franklin, and Van Aerde, were calibrated using field data from St. Louis, Missouri. The Van Aerde model fitted the field data the best as compared to the other two models. Using the Van Aerde model-generated speed–flow curves, it was found that the capacity for bridge-related activities varied from 1,416 vehicles per hour per lane (vphpl) to 1,656 vphpl and for pavement-related activities from 1,120 vphpl to 1,728 vphpl. The capacity reduction factor for different work activities was found to be in the range 0.68 to 0.95, whereas the free-flow speed reduction factor was found to be in the range 0.78 to 1.0. The methodology proposed in this paper contributes toward the development of practitioner guidance and incorporation of work activity effects into traffic impact assessment tools.

All types of work zones have some form of adverse impact on traffic. The magnitude of the impact, however, varies from one work zone to another based on a variety of factors. These factors include geographical differences, demand differences (daily traffic volumes, hourly distribution of traffic), type of work activity, capacity reductions (lane drops, narrow lanes, lack of shoulder), and lane closure configurations (*1–3*). Existing traffic impact analysis tools are capable of incorporating the demand and geometric effects of work zones. However, there is limited understanding of how work activity type influences traffic operations in work zones. This paper takes a step toward understanding the relationship between work activity type and traffic flow characteristics for freeway work zones. The work activity effects are explored using three different traffic flow measures: (1) traffic speed versus flow curves; (2) capacity reduction factors; and (3) free-flow speed reduction factors. The numerous traffic impact analysis tools, such as Quick Zone and custom spreadsheets, rely on hourly traffic volumes and capacity values (normal conditions and restricted conditions) as input to determine queue length and delay in work zones. These analytical tools incorporated the Transportation Research Board's *Highway Capacity Manual* methodology (HCM 2010) for estimating delay and queue length. The HCM 2010 (*4*) version does not have speed–flow curves specifically developed for work zone conditions, let alone different work activities. The recent version of the manual HCM 2016 (*5*) does include an extension to the methodology for development of speed–flow curves. The new version includes adjustment factors for free-flow speed and capacity of work zones based on various geometric factors, but it does not include specific guidance on how work activity effects can be incorporated. By addressing this gap in existing knowledge, this research helps to develop the future guidance for the HCM methodology for work zones.

[1]Department of Civil and Environmental Engineering, University of Missouri—Columbia, Columbia, MO

Corresponding Author:
Address correspondence to Praveen Edara: EdaraP@missouri.edu

Literature Review

In some previous studies, work zone capacity adjustment factors were developed to account for the presence of work zone activity. Al-Kaisy and Hall developed a capacity model based on data from six freeway work zones in Ontario, Canada (*1*). Several traffic, geometric, and environmental factors were considered for developing the model. Elefteriadou et al. used simulation to develop capacity models for different lane closure configurations (*6*). Work intensity was taken into account through a rubbernecking factor, which was based on another study (*1*). Krammes and Lopez developed new capacity values for short-term freeway work zone lane closures using data from 33 freeway work zones (*3*). The study recommended a base work zone capacity value of 1,600 passenger cars per hour per lane (pcphpl). The authors suggested that this value should be adjusted for heavy vehicles, presence of ramps, and type and intensity of work activity.

The effects of work zone activity on work zone capacity have been investigated by several researchers through the use of qualitative activity level variables. For example, work zone capacity values were developed by Dixon et al. based on data from 24 freeway work zones in North Carolina (*7*). Work zone intensity was categorized as heavy, moderate, or light. In a study by Jiang, traffic flow characteristics for freeway work zones were investigated utilizing a sample of four Interstate work zones in Indiana (*8*). Work zone intensity was characterized by three levels: medium, nonadjacent, and high. The mean work zone capacities for the three levels of work intensity were found to be statistically equal. Through research undertaken by Kim et al., a multiple regression model to estimate work zone capacity was developed using data collected from 12 work zone sites (*9*). Correlation analysis and linear regression were utilized by Li et al. to develop a model for work zone capacity (*10*). Intensity of work activity was investigated through the use of dummy variables for medium intensity and heavy intensity. The dummy variable for medium intensity was found to be a significant variable in the model. In a study by Zheng et al., several different work zone capacity prediction models were compared, including the HCM model, two multilinear regression models, and a fuzzy logic based artificial neural network model (*11*). The study found that the neuro-fuzzy model was the most accurate. In summary, work intensity was considered in one of two ways: an adjustment factor applied to the capacity or a dummy variable representing the work zone intensity level.

Several studies have investigated work zone capacity at specific sites, but have not investigated the effects of work zone activity on capacity. Data from two lane closures at the same construction site in Toronto, Ontario were used by Al-Kaisy et al. to investigate freeway capacity (*12*). Although there was significant variation in the freeway capacity over four days of data collection, the values developed were relatively close to the values predicted by the HCM 2010. In a study by Venugopal and Tarko, capacity models for rural freeways were developed based on data from a 15-mile interstate work zone in Indiana (*13*). The study found that rainfall, heavy vehicles, police presence, and the presence of the Indiana Lane Merge System (ILMS), an innovative traffic control system, reduced the work zone capacity.

In other related research, various aspects of work zone capacity were investigated, but the effects of work zone activity on work zone capacity were not quantified. Using Missouri field data, three different methods for computing work zone capacity were compared by Edara et al. (*14*): maximum sustained flow, rescaled cumulative flow curves, and 85th percentile flow. The most conservative capacity estimates were found using the queue discharge flow values. In another study by Chatterjee et al. (*15*), a method was developed to determine VISSIM, a microscopic simulation software, driver behavior parameters that correspond with work zone capacities. Two merging cases in an early-merge system were studied: two-to-one lane merge and three-to-two lane merge. Two car-following model parameters and one lane-changing parameter were estimated.

As discussed, most of the literature estimated work zone capacity for various geometric and traffic factors, but activity-specific capacity estimation and speed–flow curves for work zones have rarely been studied. This research developed speed–flow curves using real-world traffic flow and work activity data. For the development of speed–flow curves, three steady state traffic flow models, Gipps, Newell–Franklin, and Van Aerde (*16–18*), were calibrated using field data. The models were calibrated for all selected study locations with different work zone activity types, and statistical goodness of fit measures were utilized to determine the best model. In addition, speed–flow curves were also developed for non–work zone days (for all work zone sites) to estimate the capacity and free-flow reduction factors attributable to different work activities.

Data Collection

The impact of different work activities and lane closure configurations on traffic was studied using work zone sites from the St. Louis region in Missouri. The St. Louis Metropolitan Statistical Area is the eighteenth largest in the United States and has significant traffic demand and work zone activities (*19*). The traffic flow data for Interstate highways were collected using a combination of sensor and probe data. The University of Missouri, with the permission of Missouri Department of Transportation (MoDOT), has access to real-time traffic data throughout the state of Missouri. Probe-based speed and travel time data are available via the Regional Integrated Transportation Information System (RITIS) and Nokia HERE. Point-based sensor data of traffic flow variables are also available for different regions in the state based on sensor coverage. Three types of traffic data were collected per lane from each traffic sensor at 30-s intervals: traffic volume, speed, and occupancy.

The work zone locations selected for this study were identified from the bimonthly Regional Mobility Report published by MoDOT (*20*). The report records information from events, such as work zones and accidents, which cause significant impact on mobility. Significant impact is necessary to obtain the congested region of the speed–flow curve. The work zones in the report are classified as major, moderate, and minor, based on travel times. The report also includes information about the route, location, date, and type of work for the work zones with major and moderate impacts. The mobility reports were reviewed from January 2015 to August 2016, and 10 work zone locations were identified with major and moderate travel time impact. To discuss lane closures configurations concisely, the following convention is used: (number of total lanes, number of open lanes). For example, a location with one of the three total lanes closed because of work activity is described as a (3,2) configuration. In this study, six of the 10 work zones belong to (3,2) lane closure configurations whereas the remaining four are (2,1) and (4,2) configurations with two for each configuration.

After identification of the work zone time, the 30-s speed and flow data from the detectors were aggregated into 5-min intervals. Lorenz and Elefteriadou (*21*) found that for longer time periods (5 or 15 min), the breakdown probability at a higher flow rate is substantially higher whereas for shorter periods (1 min) the probability is low for all range of flows (low and high). Therefore, a 5-min interval was thought to be reasonable for this aggregation to capture measures of traffic flow characteristics.

Estimation of Traffic Flow Characteristic Parameters

The estimation of traffic flow parameters such as capacity, free-flow speed and jam density were necessary to calibrate macroscopic traffic flow models. The calibration was conducted for each location for both work zone and non–work zone days. The procedures that were adopted to estimate these parameters are described in this section.

Free-Flow Speed

The most accurate method for estimating segment free-flow speed is to measure speed in the field during low flow conditions [under 800 vehicles per hour per lane (veh/h/ln), HCM 2010]. Hou et al (*22*) used the 85th percentile speed as a measure to evaluate the effectiveness of posted speed limits for short-term Interstate work zones. In this study, the 85th percentile speed of traffic stream under low flow conditions, that is, under 800 veh/h/ln, was considered as the free-flow speed.

Capacity and Speed at Capacity

In much of the existing literature, capacity is defined based on pre-breakdown flow or queue discharge flow. The breakdown conditions are identified with significant drop in speed, followed by low speed sustained for some time, and fluctuating traffic flow rate (*2, 8, 23*). Some others have defined capacity as queue discharge flow (*1, 12, 24*), that is, the highest flow measured after the flow break down. Elefteriadou and Lertworawanich (*25*) found that the numerical value of pre-breakdown flow is almost always lower than the discharge flow before recovery to noncongested conditions.

To identify capacity in this study, speed and flow data were plotted in the order of time. It was observed that capacity could occur either at breakdown or before recovery to noncongested conditions. In most cases, the pre-breakdown flow was found to be higher. Thus, to prevent underestimation, the higher value between pre-breakdown flow and discharge flow before recovery was defined as capacity. The identification of capacity values is illustrated in Figure 1, *a* and *b* with time series plots for two selected work zone locations. Figure 1*a* shows the scenario when pre-breakdown flow is higher than recovery flow, whereas the opposite scenario is shown in Figure 1*b*. Speed at capacity can also be identified from these time series plots. Speed at capacity separates speed–flow observations into uncongested and congested conditions.

Jam Density

After identification of capacity and speed at capacity, field observations were split into uncongested and congested conditions. The fundamental traffic flow diagram of flow versus density is shown in Figure 1*c*. As recommended in prior research (*26, 27*), linear regression was used to create best fit lines for the uncongested and congested regions of the fundamental diagram. Jam density was estimated from the regression line for the congested state, by extending it to the x-axis. The solution of the x-axis and the regression line represents jam density for the location.

Calibration of Traffic Stream Models with Sensor Data

Over the years, different approaches, such as empirical studies, hydrodynamic analogies, and car following theories, have been used to develop speed–flow curves. Traffic flow models can be broadly classified based on their functional forms. Single regime functions can represent different traffic states with a continuous function, whereas multiregime functions address different traffic states with different equations. In this study, two single regime models, Van Aerde and modified Newell–Franklin, and a two-regime model by Gipps were calibrated for work zone conditions (*28–32*). A brief description of the models and key calibration parameters are presented next.

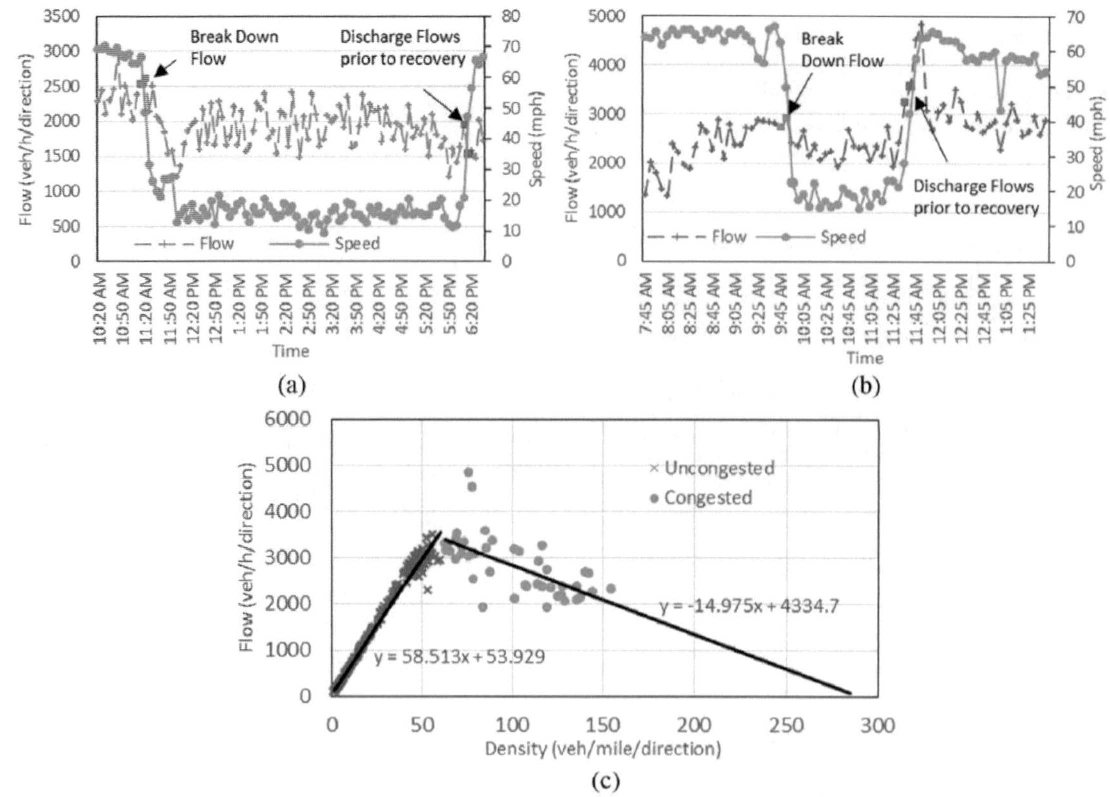

Figure 1. (a) Capacity identification when breakdown flow is higher than recovery flow, (b) capacity identification when breakdown flow is lower than recovery flow, and (c) jam density estimation from the flow–density diagram.

Van Aerde Model

The Van Aerde model is a four-parameter car following model that combines the Pipes and Greenshields models (*33*). The generalized form proposed by Van Aerde is

$$h_n = c_1 + c_3 u_n + \frac{c_2}{u_f - u_n} \tag{1}$$

where h_n is the spacing (km) of vehicle n, u_n is the speed of vehicle n (km/h), u_f is the facility free-flow speed (km/h), c_1 is the fixed spacing constant (km), c_2 is the variable spacing constant (km^2/h), and c_3 is the variable spacing constant (h).

Rakha (*30*) developed a procedure for the calibration of the four input parameters using loop detector data. The macroscopic flow and speed relation derived from the model is

$$q = \frac{u}{c_1 + \frac{c_2}{u_f - u} + c_3 u} \tag{2}$$

where q is the flow rate of the traffic stream and u is the speed of the traffic stream. The inclusion of four parameters in this model as input provides a higher level of flexibility in capturing various driver and geometric characteristics.

Gipps Model

The Gipps model is a two-regime model with an assumption that stream speed is insensitive to density within the uncongested regime. The functional form of the model (*16*), including safety margin to driver's reaction time, is

$$u_n(t+T) = 3.6\left[-bT + \sqrt{b^2T^2 + b\left\{2\left[s_n(t) - L_{n-1}\right] - \frac{u_n(t)}{3.6}T + \frac{u_{n-1}t^2}{3.6^2 \times b'}\right\}}\right] \tag{3}$$

where b and b' are the deceleration parameters of vehicle n (m/s^2), L_{n-1} is the effective length of the vehicle (m), $s_n(t)$ is

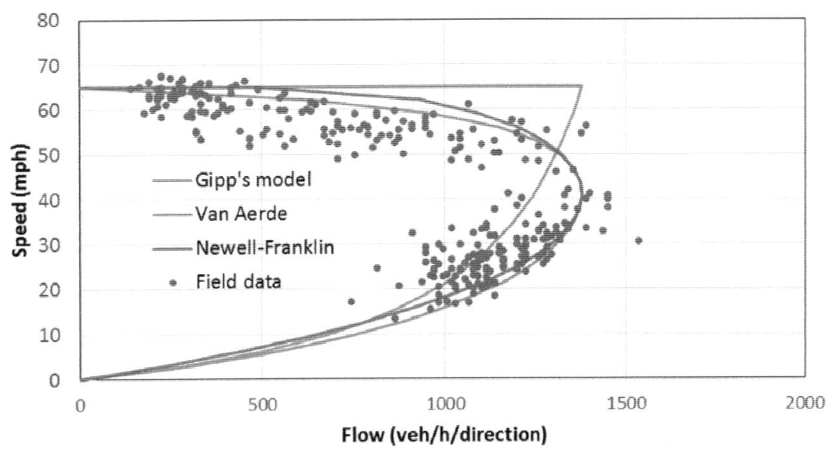

Figure 2. Example of Van Aerde, Gipps ,and Newell–Franklin models for one site.

the spacing between vehicles n and $n-1$ at time t (m), $u_{n-1}(t)$ is the speed of the preceding vehicle (km/h), and T is the driver's reaction time (s).

Rakha et al (*32*) showed a macroscopic speed–flow relation based on the Gipps car following model as

$$q = \frac{1000u}{s_j + \frac{1}{2.4}Tu + \frac{1}{25.92b}\left(1 - \frac{b}{b'}\right)u^2}. \quad (4)$$

Because of the lack of observed deceleration rates, b and b' were assumed to be equal in this study.

Re-specified Newell–Franklin Model

Li (*17*) re-specified the traditional Newell–Franklin model (*34*) by replacing jam density and kinematic wave speed with capacity and speed at capacity. The proposed generalized form is illustrated below:

$$q = q_o\left(\frac{u}{u_c}\right)\left(1 - \frac{1}{\beta}\ln\frac{u_f - u}{u_f - u_c}\right)^{-1} \quad (5)$$

where $\beta = \frac{u_c}{(u_f - u_c)}$, $q_o$is capacity, u_c is speed at capacity, and u_f is free-flow speed.

The model follows the standard practice of using capacity, speed at capacity, and free-flow speed specified by HCM for developing speed–flow curves. All three models were calibrated with work zone data. Direction-wise plots include traffic on all open lanes. Figure 2 provides an example of traffic stream models fitted to direction-wise work zone data for one site. Similar plots were generated for other sites. The details of work zone locations along with lane closure configurations, work activities, and estimated traffic flow characteristic parameters are listed in Table 1. To determine the best fit model from Van Aerde, Gipps, and Newell–Franklin models, mean absolute percentage error (MAPE) and root mean square error (RMSE) are used as goodness of fit measures. These measures are also reported in Table 1.

Goodness-of-fit measures show that the Van Aerde model was the best fit for most of the study locations (MAPE<20% for all locations). The Gipps model and Newell–Franklin model were good fits for a few locations, but experienced higher errors at other locations. The Gipps model provided a good fit for locations where the traffic stream speed in the uncongested regime remains unchanged until capacity, whereas the Newell–Franklin model provided a good fit for locations where the speed varies with flow in the uncongested regime. As the Van Aerde model fits well with the field data for all traffic scenarios, it was chosen as the best model for the remaining analysis.

Comparison of Work Zone and Non–Work Zone Speed–Flow Curves

Reduction in capacity and free-flow speed from normal conditions attributable to different activities in work zones were estimated by comparing the corresponding speed–flow curves. With regard to magnitude, the comparison of these curves does not represent the reduction in capacity and free-flow speed because of work zone activities as the curves include the loss of closed lane capacity. Therefore, for comparison purposes, the speed–flow plots by direction were scaled down to plots by lane by dividing all flow values by the number of open lanes. The speed–flow curves for normal

Table 1. Traffic Flow Parameters and Goodness-of-Fit Statistics

			Traffic characteristics				RMSE (MAPE)		
Location and Work Activity	Lane closure configuration[a]	Site number	Capacity (veh/h/direction)	FFS (mph)	Speed at capacity (mph)	Jam density (veh/mile/direction)	Van Aerde	Gipps	Newell Franklin
Eastbound I-64 at Boone Bridge (work—restriping)	(3,2)	(a)	3414	65.5	57.53	280	5.02 (7.85%)	6.72 (9.03%)	7.40 (12.26%)
Eastbound I-64 at Boone Bridge (work—bridge work)	(3,2)	(b)	3172	59.4	53	430	4.72 (9.25%)	5.71 (10.83%)	10.03 (21.19%)
Eastbound I-64 from Chesterfield Parkway to I-270 (work —paving operations)	(3,2)	(c)	3540	66.3	60	475	4.43 (6.41%)	5.45 (7.32%)	8.77 (10.65%)
Eastbound I-70 at Foristell (work—pavement repair)[b]	(2,1)	(d)	1380	65	41	115	7.14 (14.69%)	9.09 (19.64%)	6.46 (12.25%)
Westbound I-70 at Foristell (work—paving operations)	(2,1)	(e)	1560	67	60.5	100	7.73 (15.12%)	6.61 (14.21%)	9.64 (29.42%)
Eastbound I-270 from McDonnell Boulevard to Route 367(work—bridge work)	(4,2)	(f)	3408	55	43	520	6.74 (19.23%)	10.76 (33.21%)	9.79 (34.56%)
Westbound I-270 from I-170 to 367 (work—pavement work)	(3,2)	(g)	3720	69.6	60.04	450	6.16 (10.36%)	8.79 (14.42%)	8.91 (16.09%)
Eastbound I-44 at I-270 (Meramec Bridge) (work—bridge work)	(3,2)	(h)	3372	61	45	560	7.41 (10.85%)	7.84 (13.03%)	7.04 (10.56%)
Westbound I-44 (work—resurfacing)	(3,2)	(i)	2520	55	48.276	550	8.53 (19.01%)	9.64 (22.87%)	11.36 (30.95%)
Eastbound I-44 at Hampton (work—bridge rehabilitation)	(4,2)	(j)	3624	57.8	36	250	9.87 (19.66%)	13.72 (29.70%)	10.13 (20.19%)

Note: FFS = free-flow speed.
[a](number of total lanes, number of closed lanes).
[b]Site referred to in Figure 2.

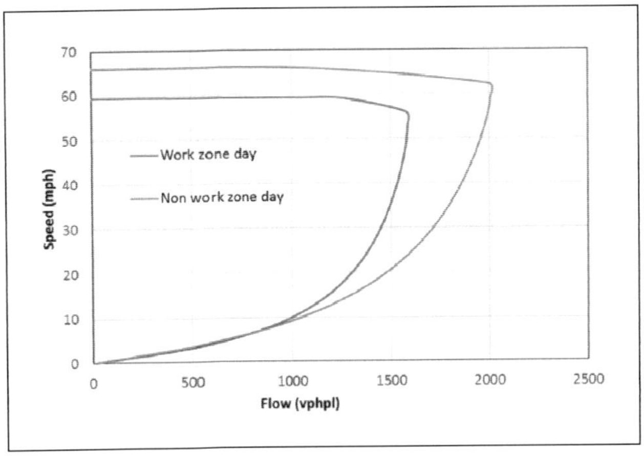

Figure 3. Sample work zone and non–work zone speed–flow curves.

days were compared with those for work zone days for all sites. Figure 3 provides an example of the two plots for one site (site (b) in Table 1).

As expected, Figure 3 shows that work zones have an adverse effect on capacity and free-flow speed. The reduction factors for free-flow speed and capacity were computed using the work zone and non–work zone values obtained from such speed–flow curves for all sites as shown in Table 2. The range of reduction factors for capacity and free-flow speed, shown in Table 2, can be extensive, depending on the nature of work and lane closure configurations. The capacity reduction factor for work zone conditions was found to be in the range 0.68 to 0.95 whereas the free-flow speed reduction factor was found to be in the range 0.78 to 1.0.

Comparison of Different Work Activity Effects

Speed–flow curves for different work activities were compared. From Table 1 and Table 2, the work activities in work zones can be broadly classified into two groups: (1) bridge-related activity, and (2) pavement-related activity. Pavement-related refers to construction and maintenance operations, whereas bridge-related includes reconstruction, joint replacement, so forth. From the MoDOT work activity list, it was found that pavement-related work includes a wide range of activities whereas the range is limited for bridge-related activities. From the literature review, it was found that work zone capacities were expressed separately for different lane closure configurations. Therefore, to be consistent with current practice, speed–flow curves developed in the present study were organized according to lane closure configuration.

In this study, (3,2) was the most common lane closure configuration observed among the work zones. Among the six (3,2) work zones, three involved bridge-related work activity and three involved pavement-related activity. Figure 4*a* shows the speed–flow curves for bridge-related activity whereas Figure 4*b* shows the curves for pavement-related activity.

Figure 4, *a* and *b* show that the variation in capacity value is much lower for bridge-related work than pavement-related work. There are only a few types of activities for bridge-related work, and this may be the reason for the low variance in capacity values (1,488 to 1,656 vphpl). In contrast, the pavement-related work has a wide range of activities, and this may be the reason for higher variation in capacity values (1,120 to 1,728 vphpl). To compare speed–flow plots of bridge- and pavement-related activities for (3,2) configuration, representative curves for both work activity groups were developed. To develop a representative curve for a group, the Van Aerde model was recalibrated for the four parameters by using average values of each parameter for all group members. The comparison is illustrated in Figure 5.

Figure 5 shows a few distinctive differences in speed–flow curves between bridge- and pavement-related activity. The free-flow speed of pavement-related activities is higher compared to bridge-related work activities. For pavement activities, the traffic stream can maintain a higher speed even at capacity conditions with negligible difference between free-flow speed and speed at capacity. In the case of bridge-related activity, however, the decline in speed is negligible until 1,300 vphpl. Once the flow continues to increase further, speed decreases significantly before capacity is reached. The decline in speed after capacity is rapid in pavement-related activities whereas the decline is gradual in the case of bridge-related activities. It appears that there is a significant difference in capacity value (160 vphpl) between bridge and pavement work. Although pavement activities have a higher free-flow speed than bridge activities, the capacity for pavement activities was found to be lower.

A similar comparison was made for (2,1) and (4,2) configurations. The representative speed–flow curves for (2,1) and (4,2) configurations with respect to work activities were developed using the methodology mentioned in the previous section. The values of capacity for the (2,1) configuration with pavement-related activities were found to be in range 1,380 to 1,560 vphpl. The capacity values for the (3,2) configuration with bridge-related activities were found to be in range 1,416 to 1,620 vphpl.

Two of the study locations consisted of (2,1) configurations with pavement-related work activities whereas there were no (2,1) sites for bridge-related activities. On the other hand, the (4,2) configuration was observed in two study locations with bridge-related activities whereas there were no such sites for pavement-related activities. Therefore, speed–flow plots for these missing activities for the two configurations were estimated using data for the (3,2) configuration. For generating the missing curves, the ratios of traffic parameters—free-flow speed, speed at capacity, capacity, and jam density—were estimated from the (3,2) configuration's pavement- and bridge-related activities

Table 2. Capacity and Free-Flow Speed Reduction Factors

Locations and Work Activity	Lane closure configuration[a]	Site number	Work zone: Capacity (veh/h): Directional (all lane)	Work zone: Capacity (veh/h): Lane wise (all lane/no of lanes)	Work zone: FFS (mph)	Non–work zone: Capacity (veh/h): Directional (all lane)	Non–work zone: Capacity (veh/h): Lane wise (all lane/no of lanes)	Non–work zone: FFS (mph)	Capacity reduction factor (10 = 5/8)	FFS reduction factor (11 = 6/9)
(1)	(2)	(3)	(4)	(5)	(6)	(7)	(8)	(9)	(10)	(11)
Eastbound I-64 at Boone Bridge (work—restriping)	(3,2)	(a)	3414	1707	65.5	6012	2004	66	0.85	0.99
Eastbound I-64 at Boone Bridge (work—bridge work)[b]	(3,2)	(b)	3172	1586	59.4	6012	2004	66	0.79	0.90
Eastbound I-64 from Chesterfield Parkway to I-270 (work—paving operations)	(3,2)	(c)	3540	1770	66.3	5748	1916	67.5	0.92	0.98
Eastbound I-70 at Foristell (work—pavement repair)	(2,1)	(d)	1380	1380	65	3360	1680	67	0.82	0.97
Westbound I-70 at Foristell (work—paving operations)	(2,1)	(e)	1560	1560	67	3360	1680	67	0.93	1.00
Eastbound I-270 from McDonnell Boulevard to Route 367(work—bridge work)	(4,2)	(f)	3408	1704	55	7212	1803	62.7	0.95	0.88
Westbound I-270 from I-170 to 367 (work—pavement work)	(3,2)	(g)	3720	1860	69.7	5976	1992	70.2	0.93	0.99
Eastbound I-44 at I-270 (Meramec Bridge) (work—bridge work)	(3,2)	(h)	3372	1686	61	5304	1768	70.7	0.95	0.86
Westbound I-44 (work—resurfacing)	(3,2)	(i)	2520	1260	55	5584	1861	70.5	0.68	0.78
Eastbound I-44 at Hampton (work—bridge rehabilitation)	(4,2)	(j)	3624	1812	57.8	7872	1968	62	0.92	0.93

Note: FFS = free-flow speed.
[a](number of total lanes, number of closed lanes).
[b]Site referred to in Figure 3.

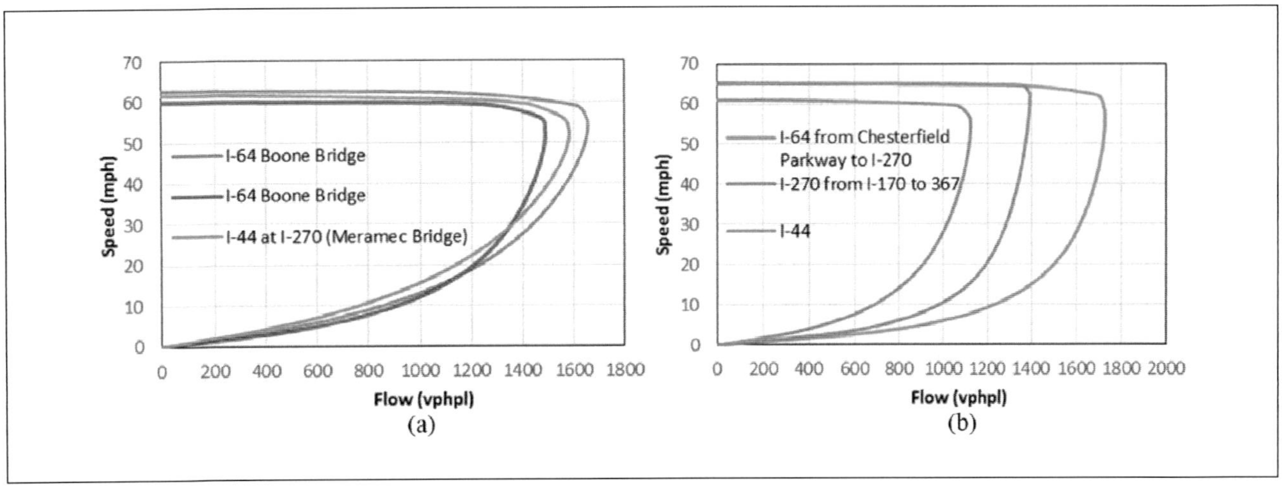

Figure 4. Speed–flow plots for (3,2) configuration: (a) bridge-related activities and (b) pavement-related activities.

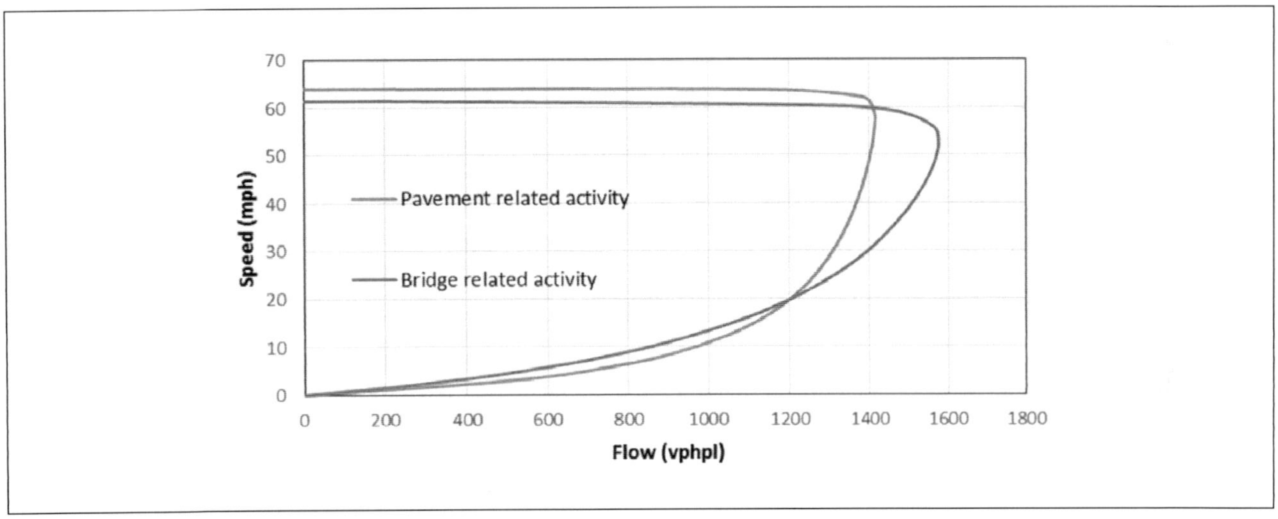

Figure 5. Comparison of speed–flow plots for bridge- and pavement-related activities for the (3, 2) configuration.

(from Figure 5). These ratios were multiplied with parameters of existing activities of the (2,1) and (4,2) configurations to estimate parameters for the missing activity type. The comparisons for the (2,1) and (4,2) configurations are illustrated in Figure 6, *a* and *b*. The dash plots in Figure 6 provide estimated values using information gathered from the (3,2) configurations. The findings of capacity by lane for both bridge- and pavement-related work activities are summarized in Table 3. From Table 3, the capacity of pavement-related work is found to be lower than for bridge-related work. The range of capacity values is wide for pavement-related work as pavement work comprises a wide range of work activity. For bridge-related activity, the range of capacity values is not as wide as work activity comprises a limited number of activities. It was also found that the (4,2) configuration has the lowest capacity, whereas the (2,1) configuration has the highest capacity among the three lane closure configurations.

Conclusions

The latest edition of HCM (2016) revamped the chapter on assessing work zone impacts. Although the 2010 edition recommended using a formula to compute work zone capacity, the 2016 edition recommends a new procedure that utilizes free-flow speed and capacity adjustment factors to determine work zone capacity. However, the new procedure does not include specific guidance on how work activity effects can be incorporated. In this paper, a methodology was developed to analyze traffic flow characteristics for various work

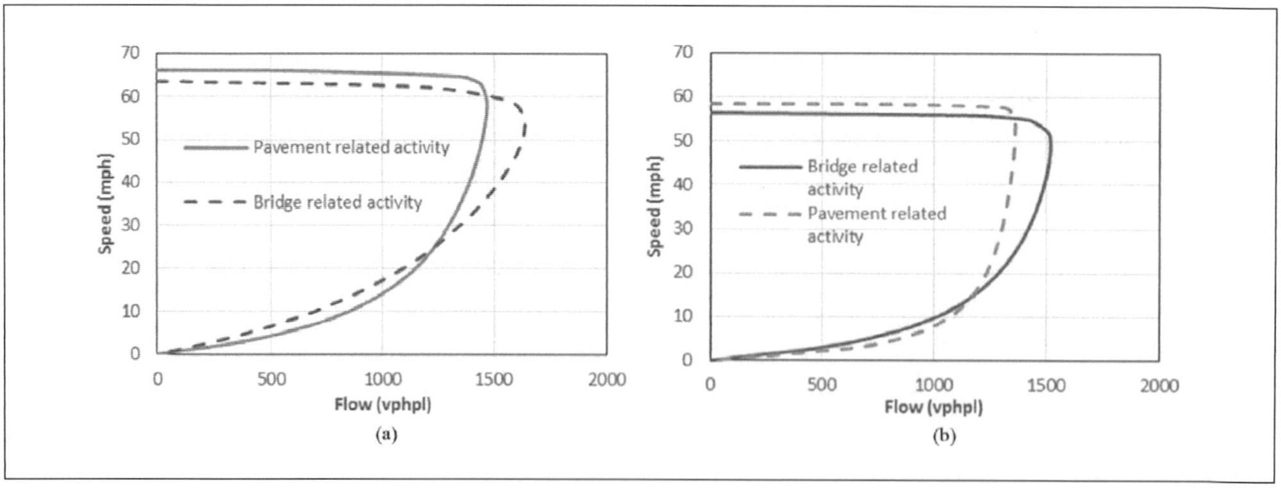

Figure 6. Comparison of speed–flow plots for bridge- and pavement-related activities for (a) the (2,1) configuration and (b) the (4,2) configuration.

Table 3. Work Zone Capacity (vphpl) Variation by Lane Closure Configuration and Type of Work

	(2,1)	(3,2)	(4,2)
Bridge Related (Range)	–	1,488–1,656	1,416–1,620
Bridge Related (Representative)	1,638.0	1575	1,518
Pavement Related (Range)	1,380–1,560	1,120–1,728	–
Pavement Related (Representative)	1,470	1,413	1,362

activities. The methodology involves developing speed–flow curves and deriving capacity and free-flow speed adjustment factors.

The following conclusions were drawn from the study:

- Higher flows could occur either at breakdown or before recovery to free-flow conditions. In most cases, the breakdown flow was found to be higher than the flow before recovery. Therefore, to prevent underestimation, a higher value between breakdown flow and discharge flow before recovery should be considered as the best estimate of capacity.
- Goodness-of-fit statistics showed that the Van Aerde model was the best fit model for most of the study locations (MAPE<20% for all locations). The Gipps model and Newell–Franklin model were good fits for a few locations, but performed poorly (MAPE>30%) in other locations. Thus, the Van Aerde model is recommended for developing speed–flow curves for various work activities.
- The capacity reduction factor for different work activities was found to be in the range 0.68 to 0.95, whereas the free-flow speed reduction factor was found to be in the range 0.78 to 1.0. This finding further emphasizes the need to customize the new HCM adjustment factors to different activity types. The factors shown in Table 2 are recommended for use by practitioners until they can develop their own adjustment factors by applying the proposed method on their jurisdictional data.
- Work activities in the work zone were broadly classified into two groups—bridge-related activities and pavement-related activities. The variation in capacity values was found to be much lower in bridge-related activities, 1,488 to 1,656 vphpl, than pavement-related activities, 1,120 to 1,728 vphpl. The higher variation in the capacity values for the pavement-related activities may be attributable to the wider range of activities that are within the scope of pavement-related work (as classified in the DOT database).

Future research should apply the proposed method to analyze additional work activities from other jurisdictions. One challenge in developing the speed–flow curves is the availability of observations representing both congested and uncongested conditions. Often, work activity may only occur during one of these two periods, thus making it challenging to fit a traffic flow model for all conditions.

A second challenge is the availability of work activity data. Work zone information is typically archived in DOT planning databases. DOTs typically archive transportation data in the form of a transportation management system database. These databases also include data pertaining to work zones. However, in the majority of the cases, the work zone information is obtained from the job descriptions contained in contracts. Even though the broad description of work and total duration mentioned in contracts may be accurate, the actual dates, type of work activity and location, and other work zone characteristics depend on weather, construction phasing, correlation with other construction activity in the area, and other parameters. Thus, relying only on planning-level data based on contracts can lead to the use of insufficient or inaccurate data to assess work activity impact. State work zone coordinators and traffic management centers (TMCs) also generate a more dynamic and short-term update of work zone schedules. For example, the TMC in St. Louis generates bi-weekly and daily schedules of work zones that it shares with work zone personnel by posting on social media and traveler information websites such as electronic alerts, Twitter, and Facebook. Because of the real-time update, these data tend to be more accurate. Unfortunately, these data are not always archived and are rarely in a format that can be easily retrieved (e.g., pdf, text, etc.) or queried later. Time and effort can be saved if DOTs can archive these real-time data and have tools available to easily retrieve activity information, that is, type of work, begin and end times, and lane closure information.

Acknowledgments

This research was conducted under the Smart Work Zone Deployment Initiative (SWZDI) and Federal Highway Administration (FHWA) Pooled Fund Study TPF-5(081). The authors thank the FHWA and the pooled fund state partners for their financial support and technical assistance. The authors thank Dan Smith from the Missouri Department of Transportation for his work in the role of project manager and his service on the Technical Advisory Committee. The authors also extend their gratitude to Jim Connell from the Missouri Department of Transportation and Dan Sprengeler from the Iowa Department of Transportation for their service on the Technical Advisory Committee.

References

1. Al-Kaisy, A., and F. Hall. Guidelines for Estimating Capacity at Freeway Reconstruction Zones. *Journal of Transportation Engineering*, Vol. 129, No. 5, 2003, pp. 572–577.
2. Heaslip, K., A. Kondyli, D. Arguea, L. Elefteriadou, and F. Sullivan. Estimation of Freeway Work Zone Capacity Through Simulation and Field Data. *Transportation Research Record: Journal of the Transportation Research Board*, 2009. 2130: 16–24.
3. Krammes, R. A., and G. O. Lopez. Updated Capacity Values for Short-Term Freeway Work Zone Lane Closures. *Transportation Research Record: Journal of the Transportation Research Board*, 1994. 1442: 49–56.
4. *Highway Capacity Manual 2010.* Transportation Research Board of the National Academies, Washington, D.C., 2010.
5. *Highway Capacity Manual: A Guide for Mobility Analysis,* 6th ed. Transportation Research Board, Washington, D.C., 2016.
6. Elefteriadou, L., D. Arguea, A. Kondyli, and K. Heaslip. *Impact of Trucks on Arterial LOS and Freeway Work Zone Capability: Part B.* Florida DOT Report No. TRC-FDOT-54954-b-2007, 2007. Florida Department of Transportation, Tallahassee, FL.
7. Dixon, K., J. Hummer, and A. Lorscheider. Capacity for North Carolina Freeway Work Zones. *Transportation Research Record: Journal of the Transportation Research Board*, 1996. 1529: 27–34.
8. Jiang, Y. Traffic Capacity, Speed, and Queue-Discharge Rate of Indiana's Four-Lane Freeway Work Zones. *Transportation Research Record: Journal of the Transportation Research Board*, 1999. 1657: 10–17.
9. Kim, T., D. J. Lovell, and J. Paracha. *A New Methodology to Estimate Capacity for Freeway Work Zones.* Presented at 80th Annual Meeting of the Transportation Research Board, Washington, D.C., 2001.
10. Li, P., L. Tao, D. Yang, and D. Wu. Influencing Factor Analysis of Freeway Work Zone Capacity and Its Estimation Model. *Proc., 4th International Conference on Transportation Engineering*, Chengdu, China, 2013, pp. 1552–1557.
11. Zheng, N., A. Hegyi, S. Hoogendoorn, H. Van Zuylen, and D. Peters. A Comparison of Freeway Work Zone Capacity Prediction Models. *Procedia-Social and Behavioral Sciences*, Vol. 16, 2011, pp. 419–429.
12. Al-Kaisy, A., M. Zhou, and F. Hall. New Insights into Freeway Capacity at Work Zones: Empirical Case Study. *Transportation Research Record: Journal of the Transportation Research Board*, 2000. 1710: 154–160.
13. Venugopal, S., and A. Tarko. *Investigation of Factors Affecting Capacity at Rural Freeway Work Zones.* Presented at 80th Annual Meeting of the Transportation Research Board, Washington, D.C., 2001.
14. Edara, P., J. Kianfar, and C. Sun. Analytical Methods for Deriving Work Zone Capacities from Field Data. *Journal of Transportation Engineering*, Vol. 138, No. 6, 2012, pp. 809–818.
15. Chatterjee, I., P. Edara, S. Menneni, and C. Sun. Replication of Work Zone Capacity Values in a Simulation Model. *Transportation Research Record: Journal of the Transportation Research Board*, 2009. 2130: 138–148.
16. Gipps, P. G. A Behavioural Car-Following Model for Computer Simulation. *Transportation Research Part B: Methodological*, Vol. 15, No. 2, 1981, pp. 105–111.
17. Li, M. Z. A Generic Characterization of Equilibrium Speed–flow Curves. *Transportation Science*, Vol. 42, No. 2, 2008, pp. 220–235.
18. Van Aerde, M., and H. Rakha. Multivariate calibration of single regime speed–flow-density relationships [road traffic management]. In *Vehicle Navigation and Information Systems Conference, 1995. Proceedings. In conjunction with the Pacific Rim TransTech Conference. 6th International VNIS. 'A Ride into the Future,'* IEEE, New York, 1995, pp. 334–341.
19. Wilson, S., D. Plane, P. Mackun, T. Fischetti, and J. Goworowska. *Patterns of Metropolitan and Micropolitan Population Change: 2000 to 2010.* 2010 Census Special Report: C2010SR-01. U.S. Census Bureau, Washington, D.C., 2012.
20. Missouri Department of Transportation (MoDOT). 2015-2016. Bi-monthly Regional Mobility Reports. http://www.modot.org/stlouis/links/traffictopics.htm. Accessed May 2017.
21. Lorenz, M., and L. Elefteriadou. Defining Freeway Capacity as Function of Breakdown Probability. *Transportation Research Record: Journal of the Transportation Research Board*, 2001. 1776: 43–51.

22. Hou, Y., P. Edara, and C. Sun. Speed Limit Effectiveness in Short-Term Rural Interstate Work Zones. *Transportation Letters: The International Journal of Transportation Research*, Vol. 5, No. 1, 2013, pp. 8–14.
23. Dong, J., and H. Mahmassani. Flow Breakdown and Travel Time Reliability. *Transportation Research Record: Journal of the Transportation Research Board*, 2009. 2124: 203–212.
24. Benekohal, R., A. Z. Kaja-Mohideen, and M. Chitturi. Methodology for Estimating Operating Speed and Capacity in Work Zones. *Transportation Research Record: Journal of the Transportation Research Board*, 2004. 1883: 103–111.
25. Elefteriadou, L., and P. Lertworawanich. *Defining, Measuring and Estimating Freeway Capacity*. Presented at 82nd Annual Meeting of the Transportation Research Board, Washington, D.C., 2003.
26. Kianfar, J., and P. Edara. A Data Mining Approach to Creating Fundamental Traffic Flow Diagram. *Procedia-Social and Behavioral Sciences*, Vol. 104, 2013, pp. 430–439.
27. Dervisoglu, G., G. Gomes, J. Kwon, R. Horowitz, and P. Varaiya. *Automatic Calibration of the Fundamental Diagram and Empirical Observations on Capacity*. Presented at 88th Annual Meeting of the Transportation Research Board, Washington, D.C., 2009.
28. Brilon, W., and J. Lohoff. Speed–flow Models for Freeways. *Procedia-Social and Behavioral Sciences*, Vol. 16, 2011, pp. 26–36.
29. Burris, M. W., and S. Patil. *Measuring the Marginal Cost of Congestion* (No. SWUTC/08/473700-00088-1). Southwest Region University Transportation Center, 2008. Texas Transportation Institute, College Station, Texas.
30. Rakha, H. Validation of Van Aerde's Simplified Steadystate Car-Following and Traffic Stream Model. *Transportation Letters*, Vol. 1, No. 3, 2009, pp. 227–244.
31. Rakha, H., and Y. Gao. *Calibration of Steady-State Car-Following Models Using Macroscopic Loop Detector Data* (DOT-HS-811483). Virginia Tech Transportation Institute, Blacksburg, Va., 2010.
32. Rakha, H., C. Pecker, and H. Cybis. Calibration Procedure for Gipps Car-Following Model. *Transportation Research Record: Journal of the Transportation Research Board*, 2007. 1999: 115–127.
33. Rakha, H., and B. Crowther. Comparison of Greenshields, Pipes, and Van Aerde car-following and traffic stream models. *Transportation Research Record: Journal of the Transportation Research Board*, 2002. 1802: 248–262.
34. Newell, G. F. Nonlinear Effects in the Dynamics of Car Following. *Operations Research*, Vol. 9, No. 2, 1961, pp. 209–229.

The Standing Committee on Work Zone Traffic Control (AHB55) peer-reviewed this paper (18-00955).

TRR
JOURNAL OF THE TRANSPORTATION RESEARCH BOARD

Article

Transportation Research Record
2018, Vol. 2672(16) 35–45

DOI: 10.1177/0361198118794284
journals.sagepub.com/home/trr
SAGE

Use of Portable and Dynamic Variable Speed Limits in Construction Zones

Josh Van Jura[1], David Haines[2], and Andrew Gemperline[3]

Abstract

The Utah Department of Transportation (UDOT) implemented dynamic management of portable variable speed limit (PVSL) technology to reduce regulatory speed limits through an active work space (AWS). UDOT also developed and tested an intelligent system approach to alter speed limits in construction work zones. The goal of the PVSL system was to provide a portable and dynamic system that was easy for construction personnel to use to prudently reduce speeds within an AWS and make construction work zones safer for workers and the traveling public, while limiting the need to reduce speed throughout the AWS, rather than the entire construction work zone. This was achieved through temporary regulatory reductions in driver speeds within the immediate boundary of an AWS when workers were on site and exposed to the danger of errant vehicles during active construction. The system also raises speed limits when workers were not present. This PVSL system used a dynamic variable speed limit (VSL) algorithm to raise and lower the regulatory speed limits. The PVSL system also provided a queue warning algorithm that operated independent of the VSL algorithm to control messages posted on the portable variable message sign (PVMS) trailers to disseminate dynamic information to drivers. UDOT has completed 2 years of PVSL system deployment testing in four separate construction work zones to evaluate the effectiveness of the system. This paper highlights key elements that guided development of the PVSL system, along with the successful results from deployment of the system.

The Utah Department of Transportation (UDOT) maintains 5,865 centerline miles of state roadways, of which 13% are posted at 80 miles per hour (mph) and 35% at 70 mph or higher. When performing construction and/or maintenance activities on these high-speed roadways, it is often not practical or feasible to physically barrier off all construction activities from adjacent traffic to provide a safe construction work zone for workers. In addition, the traditional approach to work zone speed reduction within Utah, and many other states, is to reduce speed along the entire length of the work zone. Often these reductions, intended or not, remain in place 24 hours a day, 7 days a week.

UDOT's intention to slow driver speeds only when construction personnel are present stems from their observations of similar behavior in school zones. Traditionally, the length of speed reductions in school zones frustrated drivers, leading to them dismissing the speed limit reduction. UDOT changed the length for the speed reduction to only where children were crossing, which increased driver compliance and child safety. UDOT hopes to mimic this driver behavior in construction work zones, increasing safety for workers and the traveling public.

UDOT also recognizes the work of other state transportation departments that previously deployed intelligent transportation systems (ITS), including portable variable speed limit (PVSL) systems, in construction work zones. Two relevant projects for this study were within congested urban areas along I-44 in Missouri and I-495 (the Capital Beltway) between Springfield, Virginia, and the Virginia-Maryland state line on the Woodrow Wilson Memorial Bridge. The use of ITS equipment improved traffic operations and safety in providing real-time information to travelers, monitoring traffic conditions, and managing incidents. Documented benefits and observations found that PVSL signs within a construction work zone resulted in lower average speeds and greater speed limit compliance when compared with just using static speed limit signs.

[1]Utah Department of Transportation, Salt Lake City, UT
[2]Kimley-Horn and Associates, Inc., Mesa, AZ
[3]Avenue Consultants, Taylorsville, UT

Corresponding Author:
Address correspondence to Josh Van Jura: jvanjura@utah.gov

Building upon these past projects, UDOT engaged the Utah Association of General Contractors and the Utah Highway Patrol to further develop and evaluate the use of PVSL equipment and system technology in construction work zones. UDOT submitted a grant application to the Federal Highways Administration (FHWA)'s Center for Accelerating Innovation (CAI) and was selected as a recipient of an Accelerated Innovation Deployment (AID) Demonstration grant to fund the use of a PVSL system on four roadway construction projects, with a remit to:

- Encourage significant driver speed reductions within "active work spaces" (AWS) where construction personnel are on site.
- Adjust speed limits automatically in accordance with speed harmonization principles.
- Minimize the length of the AWS for which the significant speed reductions are applied to encourage driver compliance with the speed limit.
- Increase regulatory speed limits when workers are not present.
- Provide a portable system that is easy for construction personnel to use.

While UDOT followed the overall systems engineering process, from concept of operations through system design, procurement, and testing/validation, this paper highlights key elements that guided development of the PVSL system, along with the successful results from deployment of the system on four interstate construction projects.

PVSL System Concept

Equipment Orientation and Functionality

Figure 1 depicts the equipment for a PVSL system and illustrates where the equipment is oriented to achieve operational objectives for projects requiring a single-lane drop.

Figure 1 references two operational scenarios that were initially used in defining the parameters and limits needed to operate and evaluate the PVSL system and its performance. Both scenarios are comprised of a queue warning subsystem (shown before the taper) and the PVSL subsystem within the construction work zone.

For this study, the length of the AWS was limited to two miles to minimize the distance that the drivers would experience the most significant speed limit reduction. The AWS defines the area within the larger construction work zone that has the largest concentration of workers at risk from errant vehicles. The AWS detector provides real-time speed data used by the variable speed limit (VSL) algorithm that recommends the reduced speed limits. As illustrated, Scenario 3 operates similar to Scenario 2, with the exception that more PVSL trailers are used, resulting in fewer equipment moves for construction personnel as the AWS moves down the work zone.

In addition to the equipment depicted on Figure 1, additional components developed to support the PVSL system's functionality and encourage its acceptance and use include a web-based application (Site Control System) that manages all PVSL devices, including the following features:

- Dynamic determination of the location and order of devices along the roadway, while simultaneously knowing what recommended speed limits apply to each PVSL sign as they "leap-frog" locations to keep pace with moving construction operations (see Figure 1)
- Continual monitoring of the health of the devices, while automatically sending alerts (e.g., communications loss, low power)
- Logging device locations, posted speed limits, and times when changes occurred for regulatory enforcement purposes
- Doppler radar technology used for traffic detectors, which is the easiest and fastest way for workers to deploy and move detection trailers within a construction work zone
- Cellular communication networks used for connecting intelligent devices in the AWS to the site control system
- A portable operator interface (e.g., a phone, tablet, laptop) running a web-based application used by construction personnel to monitor system performance and override system controls, if necessary

System Intelligence and Operations

The primary objective for the PVSL system is to develop and test a smart work zone system that provides construction personnel the ability to use the system as a standard traffic control approach for roadway improvement projects. To accomplish this, the system needed to:

- Establish safeguards to limit the range of speed limits that the PVSL system could post under normal operating conditions and establish rules for posting speed limits during various known potential functional or operational failures (e.g., equipment failures, erroneous speed inputs from misaligned radar).
- Develop algorithms, thoroughly tested in an AWS, that could generate posted speed limit

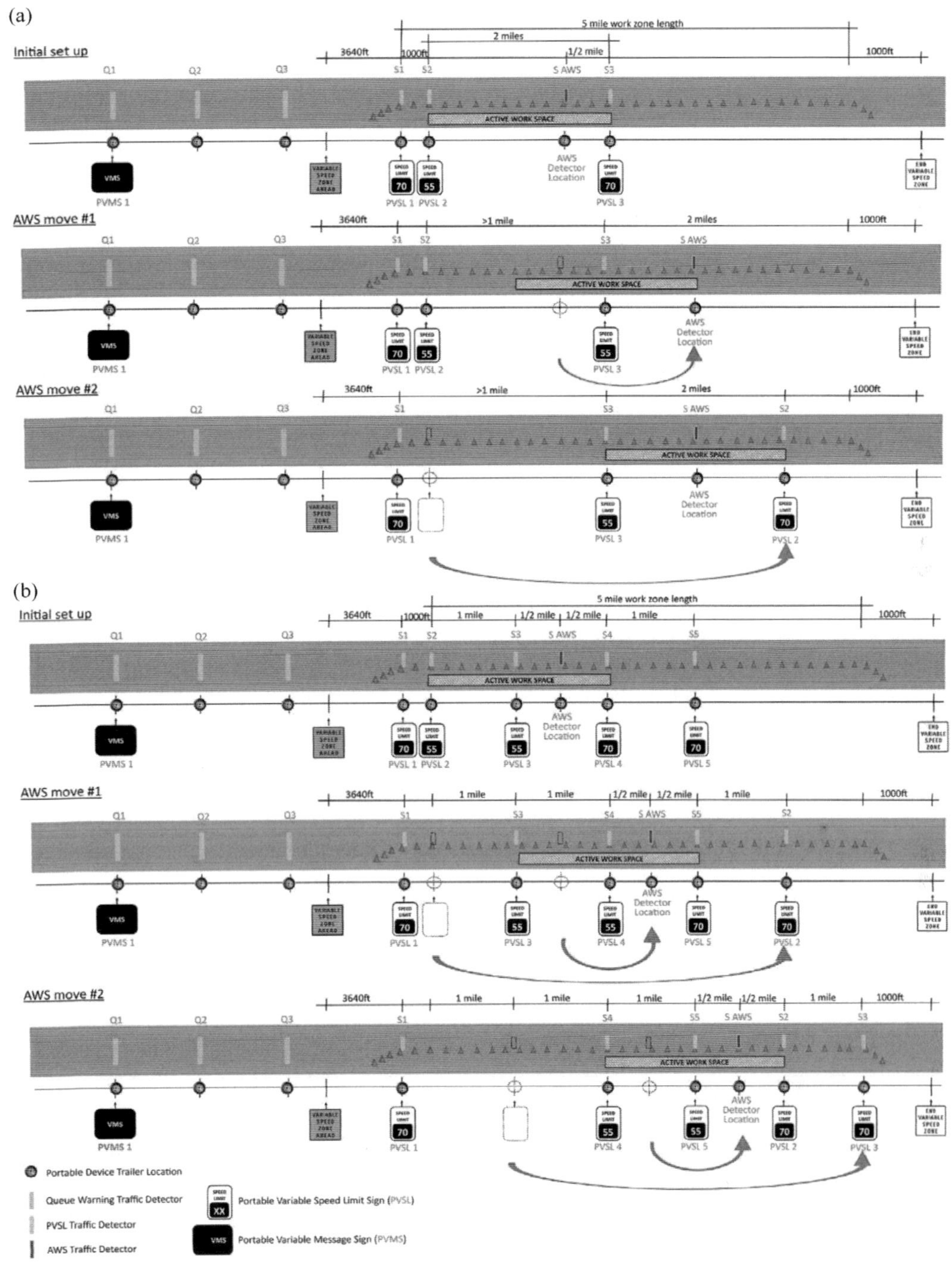

Figure 1. Operational scenario graphics: (*a*) operational deployment scenario #2; (*b*) operational deployment scenario #3.

Note: AWS = active work space; PVSL = portable variable speed limit sign; PVSL1–PVSL5 = a unique number associated with each PVSL location; PVMS = portable variable message sign; PVMS 1 = a unique number identifying the PVMS location; Q1–Q3 = a unique number associated with each queue warning traffic detector location; S1–S5 = a unique number associated with each PVSL traffic detector location.

recommendations and automatically change posted speed limits.

Safeguards Through Speed Limit Thresholds. Conditional, maximum, and minimum speed limits are established by UDOT through a Traffic Engineering Order (TEO) that applies to a specific project based on relevant work zone parameters. The TEO process is used to establish all changes in speed limits statewide and is independent of whether a PVSL system is deployed on a project. The State's traditional approach is to reduce speeds by no more than 10 mph in a construction work zone; however, this was not considered to be meeting the goal of a "significant" speed reduction to enhance worker safety. Therefore, UDOT established speed limit reduction rules for use within the PVSL system (Figure 2) to allow for greater reductions enhancing safety. This figure depicts speed reduction limits for one and two PVSL signs in advance of an AWS.

As depicted in the lower half of Figure 2, the second PVSL sign would display an equal or larger permissible speed reduction. This is done for two reasons: 1) to build driver trust regarding the legitimacy of posted speed limits due to visual confirmation of workers present shortly after they pass the second PVSL sign; and 2) to reduce the overall length of the AWS impacted by the more significant speed limit reduction.

Algorithm Development. UDOT then developed the system's operational intelligence by creating two unique algorithms (i.e., the VSL and queue warning algorithms). When combined with the speed threshold parameters and other operational limits noted in the previous section, UDOT established the basis for controlling all PVSL system devices deployed within a work zone.

The VSL algorithm raises and lowers regulatory speed limits automatically in advance of an AWS based on real-time dynamic monitoring of the speeds occurring within the AWS. The queue warning algorithm, which is independent of the VSL algorithm, dynamically changes messages posted on the portable variable message sign (PVMS) trailers located in advance of the work zone. Both VSL and queue warning algorithms are designed to generate the necessary posted speed limit and advance warning message, which reduce the effort needed to operate the system.

The process associated with the VSL algorithm is shown in Figure 3. Previous speed harmonization studies have shown that operating speeds within +/− 15 mph of an average speed are significantly safer than operating speeds outside these limits (*1–3*). The VSL algorithm was designed to make the AWS safer by applying speed harmonization principles combined with upper and lower speed limit thresholds defined by UDOT's TEO for the roadway improvement project.

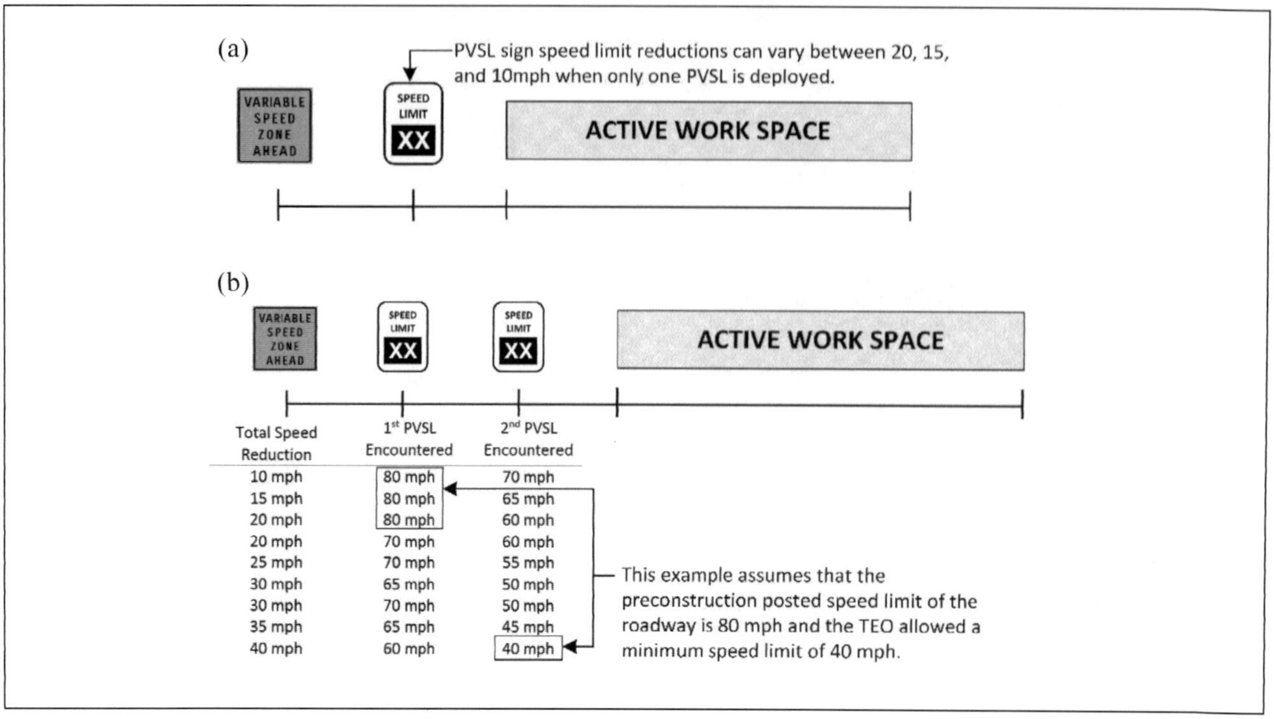

Figure 2. PVSL speed limit reduction rules: (*a*) speed limit rules for one-step speed limit reduction; (*b*) speed limit rules for two-step speed limit reduction.

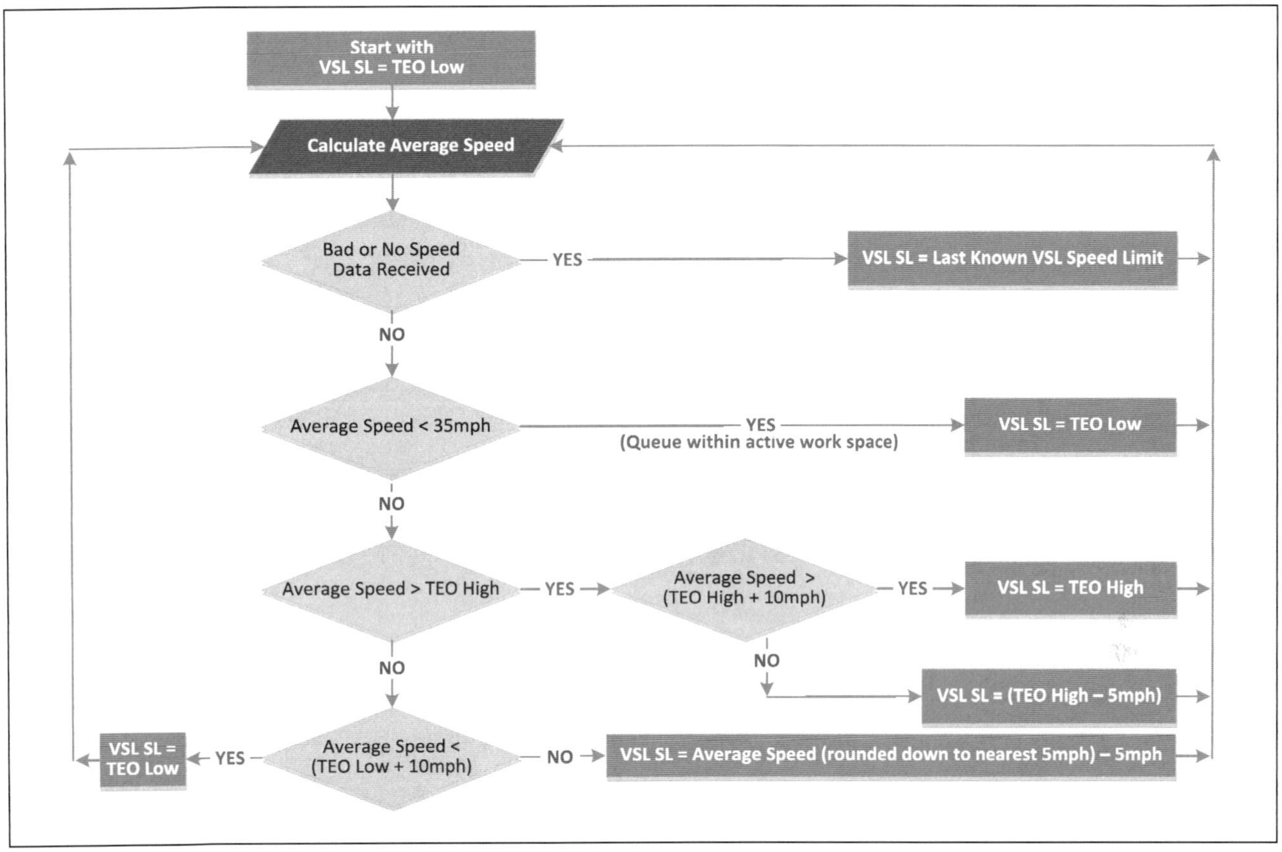

Figure 3. VSL algorithm.
Note: Average Speed = calculated average speed based on rolling 5 minutes of speed measurements in active work space; frequency of speed limit change = minimum of 15 minutes between speed limit changes; TEO = traffic engineering order; TEO high = maximum speed allowed by TEO, typically "original posted speed" or "original posted speed – 10 mph;" TEO low = the lowest speed limit allowed by TEO; VSL SL = speed limit posted on VSL sign.

The initial version of the VSL algorithm was tested using an offline, "virtual" set of PVSL signs that displayed recommended speed limits generated by the system. When the PVSL system changed a speed limit on a "virtual" sign, it concurrently sent a notification to construction personnel to input the recommended speed limit change on the corresponding "real" PVSL sign within the work zone. This required workers to log into the software from their mobile device or laptop and update the posted speed to the new recommended speed. Although the delay caused by the approval process was not ideal for testing the true performance of the system, it was a necessary step for building confidence in the system's ability to dynamically generate speed limits that are within the TEO's pre-established operational limits, including the established speed limit reduction rules (Figure 2).

The reliability of speed limit change recommendations reached a point that UDOT automated how the recommended speed changes were implemented in conjunction with the automated text messages sent to construction personnel and allowed a three-minute window to override or "not accept" the speed change.

Throughout the study, UDOT refined the VSL algorithm, starting with how the algorithm's logic was used to recommend the AWS variable speed limit by rounding down from the average speed to the nearest 10 mph increment, then subtracting 5 mph. This was followed by a variation in logic to round down from the average speed to the nearest 5 mph increment, then subtracting 10 mph. Finally, the algorithm that best adhered to speed harmonization principles rounded down to the nearest 5 mph increment and then subtracted 5 mph. These minor changes in the algorithm produced significant differences in the posted speed limit and the frequency of changes to the speed limit.

Procurement and Cost

A vendor was selected and awarded a contract covering four deployment projects for 30 days each. Equipment rental costs for the PVSL system were between \$173 and \$329 per day, depending on the deployment scenario. The queue warning system was an additional \$194 per day. In addition to the cost of renting the equipment,

Table 1. Project Year Deployment Scheduling

Project/name/posted speed/TEO LOW speed		Location	Type of construction	AADT and number of lanes	Operations scenario tested
Y1P1	Tremonton posted speed limit = 80 mph TEO LOW speed = 45 mph	I-15; Honeyville to Tremonton	Mill, paving, and bridge repair	22,000 AADT 4-lane divided highway	Scenario 2
Y2P1	Meadow posted speed limit = 80 mph TEO LOW speed = 50 mph	I-15; Baker Canyon to Meadow	Mill and paving	12,000 AADT 4-lane divided highway	Scenario 2
Y2P2	Honeyville posted speed limit = 80 mph TEO LOW speed = 50 mph	I-15; Corinne to Honeyville	Mill and paving	22,000 AADT 4-lane divided highway	Scenario 2
Y2P3	I-80 Posted speed limit = 80 mph TEO LOW speed = 50 mph	I-80; milepost 20–30	Mill and paving	7,500 AADT 4-lane divided highway	Scenario 3

system deployment budgets were required to address equipment mobilization, training, and software configuration. It was determined that a 0.5 full-time equivalent of a field person's time was needed to ensure the system transitioned in parallel with construction activities and the AWS; this was often divided between 2-3 existing employees.

Implementation

Evaluation Sites

UDOT identified four roadway construction projects that would benefit from using a PVSL system to test the system in a real-world application. UDOT selected the projects based on characteristics that helped achieve a controlled environment for system evaluation and project types that could benefit from PVSL system implementation. Based on testing results, UDOT has expanded its list of project types and characteristics where a PVSL system would benefit construction work, including:

- Projects in both rural and urban locations
- Roadways comprised of two or more lanes in each direction (divided or undivided), maintaining at least a single through lane during construction in each direction
- Roadway resurfacing and slab replacement-type construction
- Work durations of at least 30 days so that significant data may be gathered
- Roadways with enough traffic volume to measure (annual average daily traffic (AADT) volumes between 7,500 and 25,000), but not a roadway or area frequently congested
- Roadways with higher speeds (45 mph or greater)
- Roadways that are relatively flat and straight (simple geometries with minimal curves and elevation changes)

UDOT's initial test project is referred to as Year One, Project One (Y1P1). In the second year, UDOT evaluated three additional projects, which are referred to as Y2P1, Y2P2, and Y2P3, respectively. Table 1 lists all four project evaluation sites.

Base Data Collection

Baseline data was collected during the first week of each project to compare existing conditions with the results of PVSL system deployment. During this period, UDOT collected vehicle speeds and counts using the AWS detection trailer placed within the AWS. Static roadside signs were used to reduce speed limits in advance of the AWS.

Deployed System Data Collection

In the weeks following the base data collection period, UDOT deployed the PVSL trailers and detectors at locations depicted in Figure 1 for the respective scenarios. UDOT collected the same vehicle count and average vehicle speed data collected during the base data collection period during the full PVSL system deployment period. UDOT did not collect speed and count data during observed holidays, times of adverse weather, or when work was not being performed.

Y1P1: Tremonton

Figure 4 represents 12 hours of results that show a consistent and significant reduction of speeds within the AWS. The observed five-minute rolling average speeds ranged

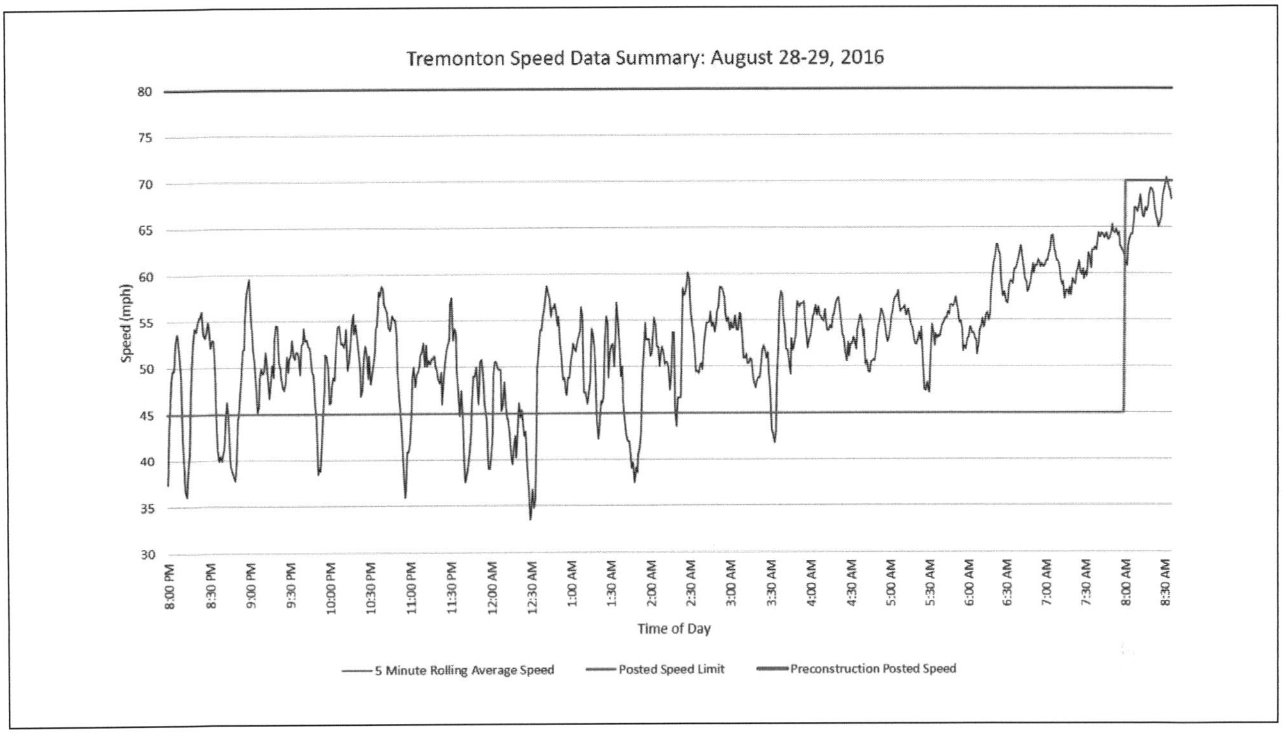

Figure 4. PVSL posted speeds compared with average speeds in the AWS.

10–15 mph higher than the reduced posted speed limit, and overall average speeds near the AWS were 20 + mph lower than the original roadway posted speed limit.

Of note, this project relied on construction personnel to manually change the speed limits based on system recommendations, as discussed in the "Algorithm Development" section. As shown in Figure 4, once the speed limit was reduced, there was no further adjustment to posted speeds for this period. Overall, construction personnel were reluctant to implement system recommendations. UDOT considered this reluctance as one of the major reasons Year Two (Y2) projects implemented automated speed changes.

This project proved how easy it was to deploy and move the system in parallel with construction work. It took less than five minutes for a single person to set and deploy a PVSL trailer for operation. The project also validated the reliability of VSL and queue warning algorithms to generate speed limits and advance-warning messages that were consistent with the rules and operational limits established by UDOT's TEO for the project.

One lesson learned was how construction equipment operating at a low speed blocked and interfered with detection of traffic. The negative effect was the system generated many recommendations. The artificially low results led UDOT to evaluate five-minute rolling average speeds instead of one-minute average speeds. Additionally, UDOT conducted sensitivity testing of the algorithm logic to determine the recommended AWS variable speed limit (e.g., "VSL SL = Average Speed (rounded down to nearest #mph) – #mph") to reduce the overwhelming frequency of recommendations sent to construction personnel.

Y2P1: Meadow

Figure 5 represents a sample set of data for the rolling five-minute average speed over one production day. Similar to Y1P1, significant reductions in speed were reached as the observed five-minute rolling average speed ranged 10–15 mph higher than the reduced posted speed and overall average speeds near the AWS were 10–20 mph lower than the original roadway posted speed limit.

The system logic has up to an eight-minute lag (identified by purple lines/arrows) before a posted speed limit changes automatically. This lag is an intentional system safeguard and accounts for the five minutes of data used to compute a rolling average speed within the algorithm and the subsequent three-minute period provided for construction personnel to reject the speed limit change before the system automatically changes the speed limit. Additionally, Figure 5 demonstrates that the system is correctly adhering to the 15-minute minimum between

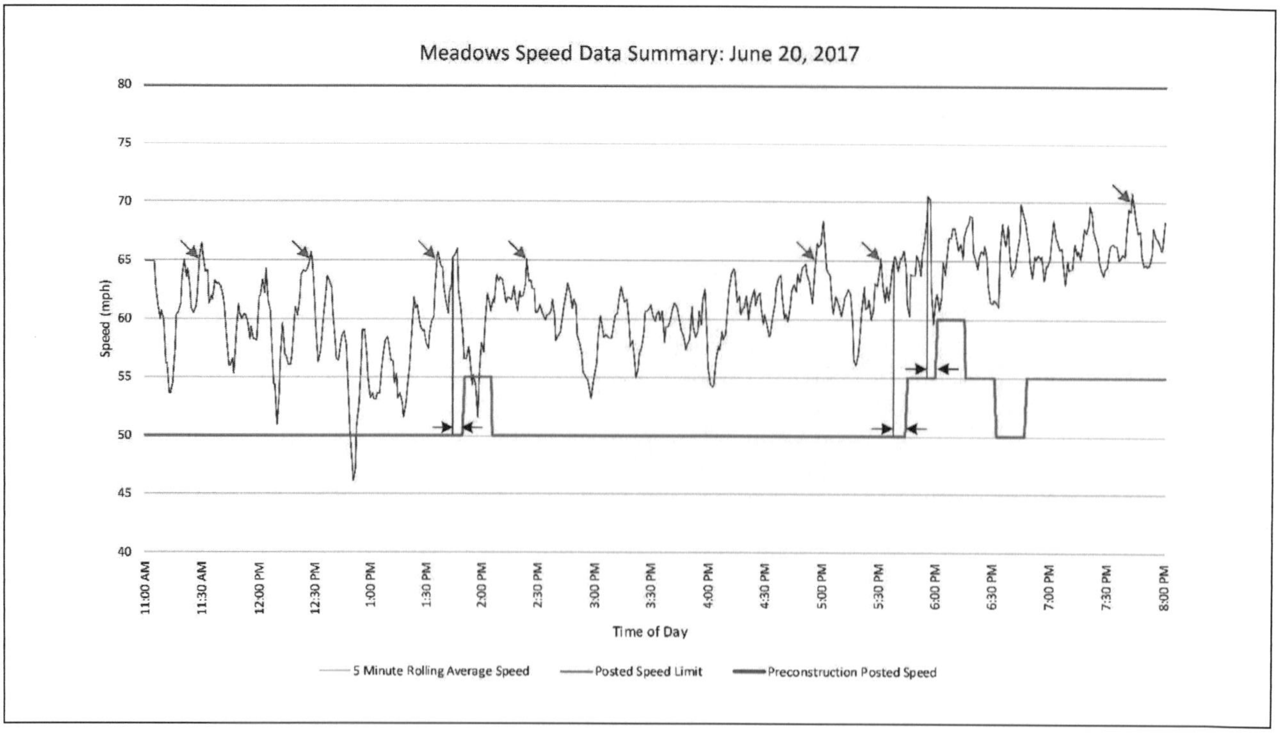

Figure 5. Example of response lag and 15 minutes between speed limit changes.

allowable speed limit changes; however, as this figure shows, the PVSL algorithm continued to require refinement, as it did not recommend speed limit changes in instances of peak five-minute rolling average speeds 15 mph over the posted speed limit as would be expected (identified by red arrows).

Y2P2: Honeyville

Y2P2 results successfully demonstrate the reliability of the functioning algorithm logic with PVSL system automation; however, it was the first deployment in which UDOT evaluated the system within an area of significant congestion that caused frequent queuing at the lane restriction taper. The congestion kept traffic at speeds below 10 mph over the posted speed limit for most of the construction work day. Under the congested conditions, the algorithm logic used to generate and recommend the AWS speed limit rounded down the rolling average speed to the nearest 5 mph, then subtracted 10 mph, resulting in the expected speed limit recommendations by the system.

While system automation simplified use for construction personnel, moving the PVSL equipment through the sequence depicted for Scenario #2 (Figure 1) as construction activities moved down the road became problematic, resulting in the posting of default speed limits (Figure 6). The AWS detector should have been the first device moved. When construction personnel did not follow this sequence, both PVSL devices posted the TEO-high 70 mph speed limit until the AWS detector was moved into its proper position. If the proper sequence of operation was followed, both PVSL devices would have posted 55 mph (the last recommended AWS speed limit) until PVSL #1 was moved into its proper position downstream of the AWS.

While Y2P2's traffic patterns were not conducive to evaluating the AWS speed reduction performance of the PVSL system, the system's queue warning operational capability successfully demonstrated automatic triggering of the queue warning subsystem, which advised travelers of slow/stopped traffic ahead. Because this was the first time the PVSL system was tested in a high volume of traffic, the results showed significantly lower actual speed conditions, unrelated to the use of the system.

Y2P3: I-80

Y2P3 was the first project that implemented Scenario 3 (Figure 1) for PVSL system operations, with six PVSL signs placed with approximately one mile of separation between each device. The project location had a history of excessive speeding during construction and non-construction times. The project's 10-mile work zone was

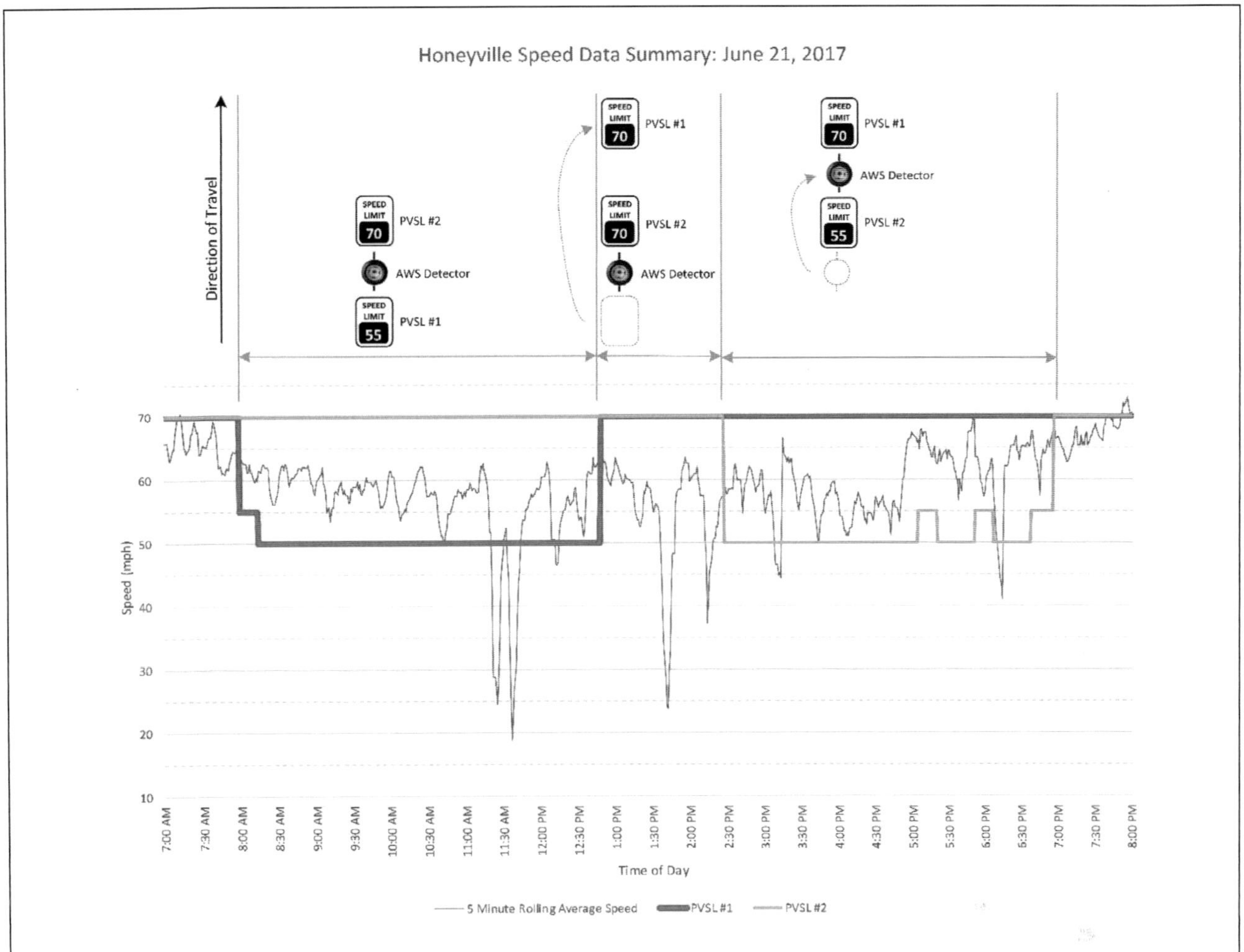

Figure 6. Scenario 2 example of equipment not moving in proper order.

an ideal candidate for demonstrating the advantages of Scenario 3. By placing six PVSL devices at one-mile intervals, the frequency of relocating each device was significantly minimized.

Figure 7 shows the observed five-minute rolling average speed ranged 10–15 mph higher than the reduced posted speed, and overall average speeds near the AWS were 15–25 mph lower than the original roadway posted speed limit. This was the greatest reduction to date showing significant AWS speed reductions and compliance with speed harmonization principles, as exhibited by the close correlation between the five-minute rolling average and the posted speed limit.

Implementing a new algorithm logic influenced the results as the recommended AWS speed limit was derived from rounding down the rolling average speed to the nearest 5 mph increment, then subtracting 5 mph. The additional speed limit changes allowed the system to be more responsive to changing traffic patterns, thereby achieving compliance with speed harmonization principles.

Figure 7 also demonstrates that the system is correctly adhering to the 15-minute minimum between allowable speed limit changes.

Another system safeguard that is illustrated in Figure 7 is what the posted speed limit should be when no traffic data is received from the AWS detection trailer, as was the case from 2:00 to 2:20 p.m. For this project, the safeguard rule was set to "Post the TEO High speed limit" if no data was received from the AWS detection trailer (i.e., the five-minute rolling average speed is zero). At the time the rule was set, it seemed to be a logical safeguard. Now that UDOT sees how this safeguard can impact traffic patterns, UDOT determined that the rule should be re-defined to "Post the last known speed limit," if no data is received from the AWS detection trailer. It is

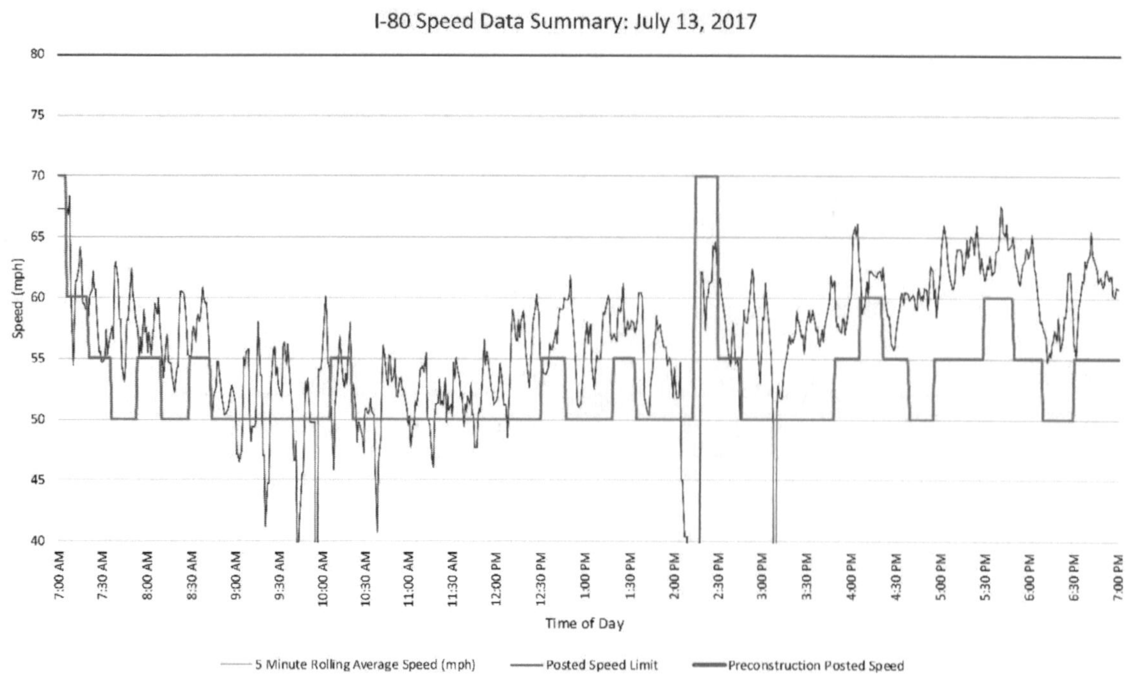

Figure 7. Scenario 3 example with algorithm logic using round down to the nearest 5 mph and subtract 5 mph.

Table 2. Speed Compliance Results

Project		Speeds < 10 mph over posted speed		10 mph < speeds < 15 mph		Speeds >= 15 mph over posted speed		Operations scenario tested
		Baseline	PVSL	Baseline	PVSL	Baseline	PVSL	
Y1P1, Tremonton	Overnight (12-hour period)	39%	41%	16%	20%	45%	39%	Scenario 2
	Daytime (12-hour period)	23%	37%	10%	15%	67%	48%	Scenario 2
Y2P1, Meadows		88%	58%	9%	29%	3%	13%	Scenario 2
Y2P2, Honeyville		N/A	60%	N/A	29%	N/A	10%	Scenario 2
Y2P3, I-80		43%	81%	42%	14%	15%	6%	Scenario 3

important to note that construction personnel were moving the AWS detection trailer during this period of time to keep pace with moving construction operations, and this was why no detection data was being received, as the sign face and radar was turned 90 degrees when transporting to reduce wind load.

Speed Compliance Results and Enhanced Safety

Table 2 summarizes the speed compliance results for all four projects.

The data for the Scenario 2 projects (Y1P1, Y2P1, and Y2P2) all exhibit excellent compliance with the speed reductions displayed on the PVSL signs, which signifies a significant reduction in operating speed. When Scenario 2 projects are compared with the baseline data, there are mixed results. Y1P1 showed improved compliance results, whereas Y2P1 showed reduced compliance results. The baseline data for the Y2P2 project was collected with a posted speed limit that was incompatible with the TEO, rendering comparison with PVSL operations debatable.

The compliance results for the Scenario 3 (Y2P3) demonstrate significant improvement in compliance compared with the associated baseline data. This was the last project deployed, so it may have benefited from changes in the algorithm variables to derive the recommended AWS posted speed using the "round down to the nearest 5 mph increment and subtract 5 mph." Another factor that may have contributed to the significant performance improvements, was the presence of more PVSL signs evenly spaced along the work zone that could have enhanced drivers' confidence in the system and/or their attention to the speed limits being posted. The system was more responsive to changing traffic patterns, thereby achieving compliance with speed harmonization principles.

Experience of Enhanced Safety

During Y1P1, while crews were paving at night with the PVSL system active and the speed reduced from 80 mph to 45 mph, a tractor trailer swerved into the barreled-off construction work zone and came to a stop within 60 feet of the paving machine. When the two individuals on the rear of the paver saw the rapidly approaching headlights, they began to scurry up and over the machine. The use of a PVSL system, in this case, was proved to save the lives of the construction personnel and perhaps the driver of the truck, which would have hit the paving machine at approximately 30–45 mph had the speed limit not been reduced. This conclusion was based on the reduced braking distance of 276 feet for a passenger car, when there is a 25-mph reduction in speed from a 70-mph speed limit, which enabled the truck driver to come to a stop before hitting the paver (*4*).

Conclusion

The following summarizes the quantitative evidence and qualitative observations of PVSL system performance.

The PVSL system functioned reliably with enough predictability and consistency in deriving recommended speed to enable automation of posting speed limits using PVSL equipment. Significant speed reductions, within the limited length of the AWS, were observed for the five-minute rolling average speed, ranging 10–15 mph higher than the reduced posted speed. Overall average speeds near the AWS were 15–25 mph lower than the original roadway posted speed limit. Significant improvement in speed compliance occurred when the algorithm logic recommended speed limits using "round down to the nearest 5 mph increment and subtract 5 mph." The close correlation between the five-minute rolling average with the posted speeds shows compliance with speed harmonization principles. Overall, the use of the PVSL system reduced the impact on drivers by limiting the length and duration the drivers had to reduce their speed and encounter the maximum speed limit reduction. The queue warning subsystem successfully demonstrated automatic triggering of the messaging to advise travelers of slow/stopped traffic ahead.

Additionally, construction personnel's perception of the PVSL system's impact on safety was that automation improved the safety of the workers because they did not have to drive to each device trailer to change messages being posted, traffic successfully slowed in the AWS, and drivers seemed to pay more attention to VSL over static signs.

Based on the positive results from the four trial projects, UDOT intends to create specifications and drawings to allow roadway contractors to bid and deploy a PVSL system. UDOT will also continue to refine best practices for appropriately selected projects in high-speed construction work zones throughout the State. The PVSL system could easily be implemented in other states to improve worker safety given that the states have legal authority to enforce VSL.

Author Contributions

The authors confirm contribution to the paper as follows: study conception and design: Josh Van Jura, David Haines, Andrew Gemperline; data collection: Josh Van Jura, David Haines, Andrew Gemperline; analysis and interpretation of results: Josh Van Jura, David Haines, Andrew Gemperline; draft manuscript preparation: Josh Van Jura, David Haines, Andrew Gemperline. All authors reviewed the results and approved the final version of the manuscript.

References

1. West, L. B. Jr., and J. W. Dunn. Accidents, Speed Deviation, and Speed Limits. *Traffic Engineering*, Vol. 41, No. 5, 1971, pp. 52–55.
2. Kloeden, C. N., A. J. McLean, V. M. Moore, and G. Ponte. *Travelling Speed and the Risk of Crash Involvement*. NHMRC Road Accident Research Unit, University of Adelaide, Adelaide, Australia, 1997.
3. Kloeden, C. N., A. J. McLean, and G. F. V. Glonek. *Reanalysis of Travelling Speed and the Risk of Crash Involvement in Adelaide South Australia*. Australian Transport Safety Bureau, Adelaide, Australia, 2002. http://casr.adelaide.edu.au/speed/vol-1.html. Accessed November 1, 2017.
4. *A Policy on Geometric Design of Highways and Streets*, 6th ed. AASHTO, Washington, D.C., 2011, pp. 3–4.

The Standing Committee on Work Zone Traffic Control (AHB55) peer-reviewed this paper (18-00585).

Article

TRR
JOURNAL OF THE TRANSPORTATION RESEARCH BOARD

Transportation Research Record
2018, Vol. 2672(16) 46–56

DOI: 10.1177/0361198118798297
journals.sagepub.com/home/trr
SAGE

Validating the Performance of the FHWA Work Zone Model Version 1.0: A Case Study along I-91 in Springfield, Massachusetts

Andrew Berthaume[1], Lauren Jackson[1], Ian Berg[1], Brian O'Donnell[2], and Christopher L. Melson[3]

Abstract

Central to the effective design of work zones is being able to understand how drivers behave as they approach and enter a work zone area. States use simulation tools in modeling freeway work zones to predict work zone impacts and to select optimal design and deployment strategies. While simple and complex microscopic models have been used over the years to analyze driver behavior, most models were not designed for application in work zones. Using data collected from an instrumented research vehicle and model components from two PhD dissertations, FHWA created the Work Zone Driver Model and programed the Work Zone Driver Model DLL v1.0, a software that could override car-following in commercial microsimulation software packages so that practitioners can better predict work zone impacts. This paper demonstrates the capabilities of the FHWA Work Zone Driver Model DLL v1.0, interfaced with VISSIM and tested on an Interstate work zone in Springfield, Massachusetts. The dynamic link library's (DLL's) performance is compared with field data collected using an instrumented research vehicle and to Weidemann 99 in VISSIM. Performance metrics were selected to align with state department of transportation work zone management efforts. Results showed acceptable performance from the DLL, as it predicted queue locations and travel speeds that were near field observations. Limitations of the DLL and interface are discussed, and opportunities for improving version 2.0 are described.

The Interstate system is comprised of approximately 223,000 lane-miles of pavement, has more than 57,000 bridges, and tens of thousands of other significant structural elements (*1*). Many of these elements are reaching their design life of 50 years and require major rehabilitation or replacement in the near future. Work zone activity has already been increasing over the past several years, and the number of work zones is projected to increase (*2*).

Freeway work zones can create significant delays. They account for nearly 24% of all non-recurring delay and 10% of total delay annually (*3*). Even during off-peak hours, work zones are a major contributor to delays, averaging 11 hours of delay per non-peak urban traveler (*4*).

To mitigate work zone impacts, states develop traffic management plans (TMPs) with targeted traffic control strategies for significant construction projects. The first step of a TMP is predicting the operational impacts of the work zone (*5*). Engineers and planners have used both microscopic and macroscopic models to predict the queues, delays, and travel times created by work zones within a degree of accuracy, giving them the foresight to design and schedule a work zone that will reduce the total delay (*6*, *7*).

Microscopic models calculate and predict the state of individual vehicles in continuous or discrete time-space, and they offer detailed descriptions of both road and traffic characteristics critical for work zone modeling. By modeling the individual reactions of vehicles, microscopic models provide a more precise look at how network elements affect traffic flow (*8*). If an accurate model were used, the aggregate delays experienced by each vehicle could provide practitioners with detailed insight about

[1]Volpe, The National Transportation Systems Center, U.S. Department of Transportation, Cambridge, MA
[2]Stinger Ghaffarian Technologies, Incorporated, Cambridge, MA
[3]Turner-Fairbank Highway Research Center, McLean, VA

Corresponding Author:
Address correspondence to Andrew Berthaume: Andrew.Berthaume@dot.gov

work zone impacts. While these tools have been used to predict work zone impacts in the past (*9*), their core algorithms were not developed to reflect the specific behaviors observable in work zone driver behavior. Before microscopic models can be used to accurately predict work zone impacts, a work zone driver behavior model needs to be developed.

FHWA developed a work zone microscopic model to simulate car-following through a work zone. The FHWA Work Zone Driver Model, which was designed to better predict vehicle responses to work zone elements, was created using field data, and synthesized from two PhD dissertations. The algorithms were calibrated and validated at a microscopic level. So that practitioners could use the model to predict work zone impacts, FHWA developed the Work Zone Driver Model v1.0, which, when interfaced with a commercial microsimulation software package, overrides the car-following logic for work zone segments. The software is implemented as a dynamic link library (DLL).

Before practitioners can use this software, it must be interfaced with a microsimulation software package and tested to validate its performance and accuracy.

The purpose of this research is to test and validate the performance of the FHWA Work Zone Driver Model DLL v1.0 through a case study, so that practitioners can use the software to better estimate the impacts of freeway work zones. Validation performance metrics were selected to align with state and local department of transportation (DOT) performance measures often used in TMPs (e.g., travel time, delay, back of queue).

For this case study, the FHWA Work Zone Driver Model v1.0 was interfaced with VISSIM 7.0 and tested along the northbound travel lanes of an instrumented work zone on I-91 in Springfield, Massachusetts. The accuracy of the FHWA Work Zone Driver Model DLL v1.0 was tested; results were compared with VISSIM's car-following model, Weidemann 99, and to field data.

Literature Review

States are interested in aspects of work zone performance and use work zone performance measures for various reasons. Hallmark et al. (*10*) summarize the various work zone performance metrics used by state DOTs. While the list is not exhaustive, it does summarize usage of some of the most popular metrics; total or average delay time was the most common, but other common metrics included queue length and location, number of vehicles that encounter a queue, and maximum per-vehicle delay. Some measures, like user satisfaction or complaint figures, were also common, but cannot be modeled with microsimulation.

Each of these metrics is useful for different purposes. For example, it has been suggested that motorists view work zones primarily in terms of delay (*11*, *12*), so states collect delay times to inform road users about potential impacts. Other metrics are used in the planning stages, since states are required to evaluate road construction projects that use federal funds for mobility impacts throughout the planning and construction process (*5*).

The choice of certain performance metrics often depends on a state's unique circumstances. Rather than suggesting that states track a large number of metrics, federal guidance recommends a limited number of well-tracked key metrics, provided there are established performance targets for each measure (*13*). For example, Michigan's thresholds for work zone performance include travel time delay of less than 10 minutes and a volume-to-capacity ratio of less than 0.80 (*11*).

Many computer models have been developed to predict work zone impacts. Research has, for example, focused on improving capacity predictions (*14*, *15*) or comparing various tools designed to predict work zone performance (*6*, *7*). These tools have had varying levels of success; for instance, Schnell et al. (*7*) noted that the models they tested consistently underestimated queue length, and in some cases were unsuitable for work zone modeling.

Simulation at the microscopic level, in particular, has recently become more common as a tool to evaluate the potential impacts of work zones. Pringle and Nikolic (*16*) cite its utility to practioners as a customizable tool, its ability to capture traffic dynamics over time, and the potential to consider and evaluate multiple underlying elements and factors. They also highlight that the visual element that microsimulation tools provide are useful in presenting results to non-technical audiences or for high-level discussion.

On the other hand, microsimulation requires calibration and validation to elicit useful results—Pringle and Nikolic (*16*) warn of the risk of "garbage in, garbage out." To encourage the proper use of microsimulation, the state of Washington has established a VISSIM protocol (*17*), which defines minimum standards that a model must meet in the calibration and validation process in order to be used in project development in the state. These include performance standards for error in speed and capacity predictions, as well as guidelines for checking model rationality and errors.

These steps are crucial to a model's performance, as reflected by the amount of literature on the topic. For example, research has been done to attempt to calibrate VISSIM's default parameters based on estimated work zone capacities (*18*, *19*). Improvements to capacity values, however, do not necessarily suggest that other estimates will also be improved. Chatterjee et al. (*19*) simulated a wide variety of potential values to develop a set of lookup tables that allow practitioners to choose calibrated parameters based on estimated work zone capacities and lane distribution. While they assume that these calibrated parameters will result in more accurate outputs when it comes to metrics like queue length and delay, such assumptions cannot be taken for granted and must be separately validated (*20*).

A common way to validate any car-following model is to compare its performance with field data other than the data with which it was calibrated. This verifies that a model does not overfit the calibration data, but still performs successfully in situations it was intended to model. This review found that, generally, such models are compared with field data by reporting the error in one or more performance metrics, expressed (for example) as percent error (*21*) or root mean square error (*22, 23*). The lower the error, the better the model is said to have performed, although Hollander and Liu (*20*) note that minor fluctuations are in the nature of traffic, so some level of error is to be expected. From this review, it was determined that a primary concern in validation is that field data are collected and analyzed using the same methods as the model data, so that variances in data collection or data processing proceedures do not influence validation results.

Conclusions from this review include the following:

1. The performance metrics of speed through the work zone, queue length and location, and travel times through the work zone would align with state and local DOT interests in work zone performance;
2. Validation efforts should directly compare performance metrics from the model with data collected in the field. Field data should be processed using the same methods as model data, for a true comparison.

Overview of FHWA Work Zone Driver Model – Version 1.0

The FHWA Work Zone Driver Model was developed to better predict car-following through freeway work zones. The model uses the framework and acceleration equations developed in two dissertations:

1. *A New Multidimensional Psycho-Physical Framework for Modeling Car-Following in a Freeway Work Zone* (*24*). A psycho-physical car-following framework that defines relative-distance (Δx) and relative-speed (Δv) regions with distinctly different driver behaviors;
2. *Modified Field Theory* (MFT) (*25*). Acceleration equations created for work zones, derived from psychology's Field Theory.

The FHWA Work Zone Driver Model v1.0 calculates car-following behaviors differently based on the segment traversed (freeways, advanced warnings, taper zones, and work zones), the level of congestion (congested vs. uncongested), and classification of the lead vehicle (passenger car, heavy vehicle, motorcycle). In the model, each unique scenario (e.g., advanced warning, congested conditions, following a passenger car) has a unique framework with its own calibration variables.

Each framework includes seven "regimes," bounding specific regions in the $\Delta x/\Delta v$ space. Figure 1 shows the regimes.

Within each regime, a combination of force equations is used to compute car-following acceleration response. Relative speed, relative position, following vehicle speed, and parameters describing driver preference (such as desired speed or safe following distance) are factored into force equations to calculate the response. The resulting acceleration/deceleration for each modeled timestep is implemented in the software.

FHWA Work Zone Model DLL v1.0

The FHWA Work Zone Model v1.0 was implemented in a DLL that, when interfaced with a microsimulation software package, overrides the microsimulator's car-following algorithms for work zone segments. This DLL is intended for use by engineers and planners to predict the impact of freeway work zones.

Case Study: I-91 Springfield, MA Work Zone

The performance of the FHWA Work Zone Model in VISSIM was tested against an instrumented work zone along Interstate 91 northbound in Springfield, Massachusetts. This three-lane, 6-mile segment included five on-ramps, seven off-ramps, and a 1 + mile work zone that reduced capacity from three lanes to two lanes. To evaluate model performance, driver behavior field data were collected using an instrumented research vehicle (IRV) during the weekday a.m. peak hour (08:00–10:00) (*26, 27*). To create and calibrate the VISSIM networks, macroscopic data were collected and summarized by Massachusetts department of transportation (MassDOT) using roadside instrumentation (*28*). This work zone was selected because it was a long-term work zone (4 + years) and the segment was heavily instrumented. Figure 2 shows the segment of I-91 examined.

Network Calibration

Volume. Traffic volumes were defined using count data found on the MassDOT Highway Transportation Data Management System website (*28*). The location of each

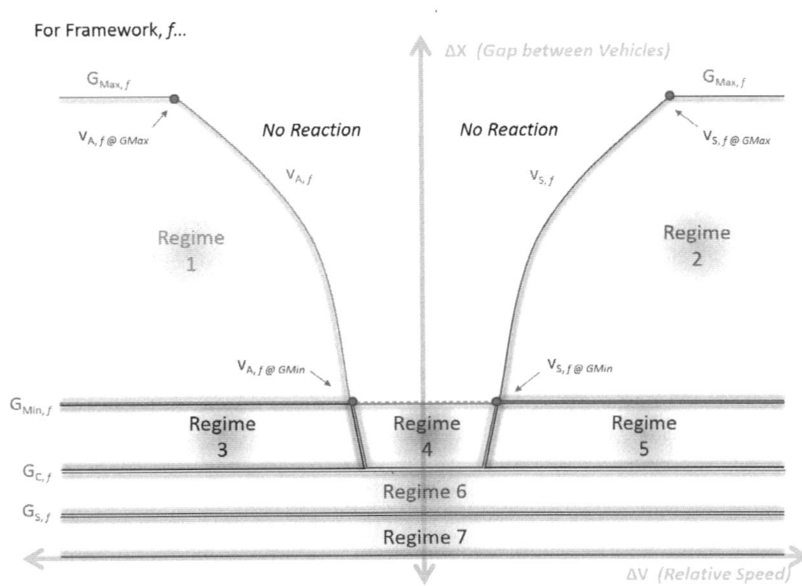

Figure 1. Generic car-following framework of the FHWA Work Zone Driver Model.

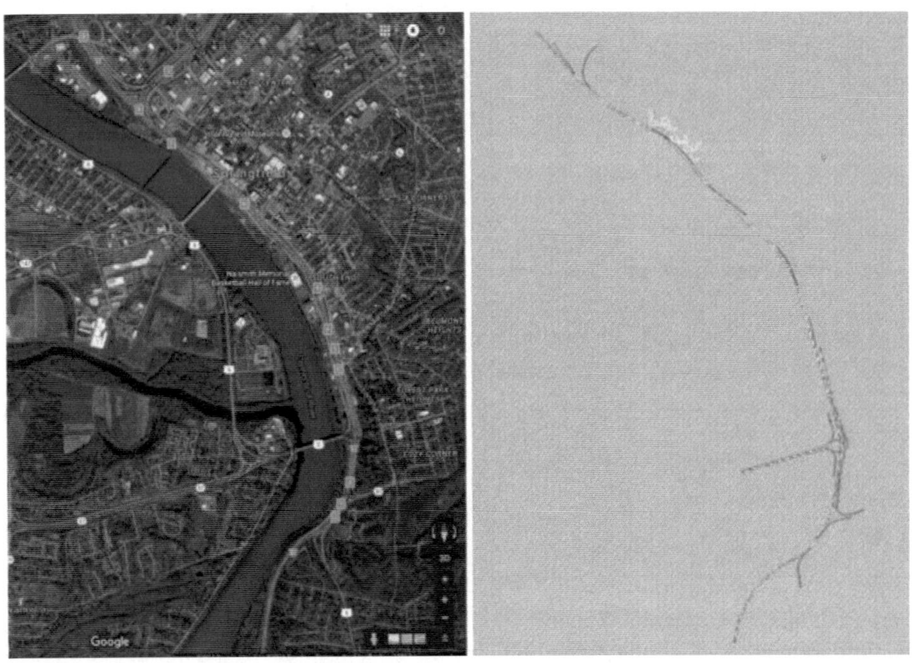

Figure 2. I-91 in Springfield, Massachusetts:satellite image (left) and VISSIM network (right).

field data collection point is shown in Figure 3. The vehicle input volumes entered into the VISSIM network are also provided in Figure 3. Volumes are representative of the field data collected July 9, 2016, between 07:30 and 10:00 a.m. The volume for Point 3 (Exit 2 on-ramp) was estimated due to lack of data.

Table 1. Desired Speed Distribution for I-91 from MassDOT

Speed bins (mph)	40–44	45–49	50–54	55–59	60–64	65–69	70–74	75–79	80–84
Midpoint speed (mph)	42	47	52	57	62	57	72	77	82
Fitted normal cumulative northbound distribution	0.0%	0.2%	1.6%	8.4%	26.7%	55.4%	81.4%	95.1%	99.2%

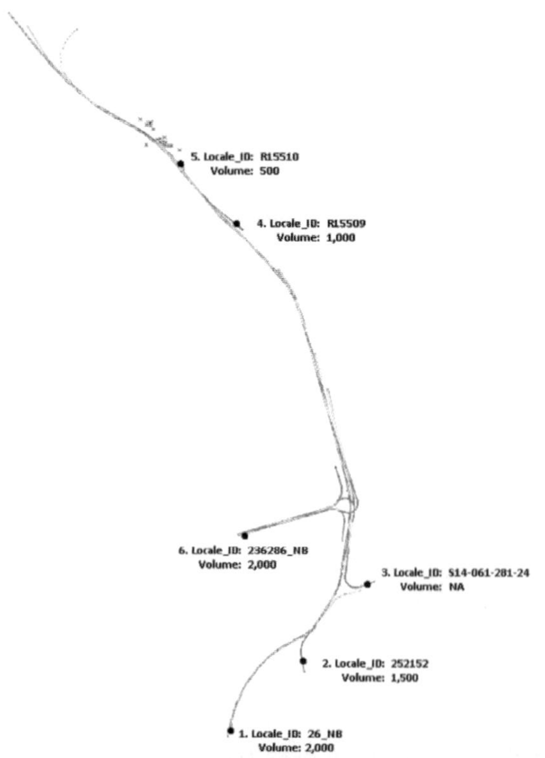

Point	Locale_ID	Road Name	Volume
1	26_NB	I-91 NB	2,000
2	252152	Exit 1 On-Ramp	1,500
3	S14-061-281-24	Exit 2 On-Ramp	1,500 (estimated)
4	R15509	Exit 3 On-Ramp	1,000
5	R15510	Exit 4 On-Ramp	500
6	236286_NB	US 5	2,000

Figure 3. MassDOT data collection points used for Interstate 91 (I-91) northbound (NB) vehicle input volumes.

Desired Speed. Speed distributions were created using historical records for I-91, published on MassDOT's Highway Transportation Data Management System site (*28*). The desired speed distribution published to this website is consistent with field data collected using the IRV. The data used for this study were collected July 7, 2016, and are listed in Table 1.

Due to limited number of speed data collection points, the data used to calibrate VISSIM's desired speed distribution were collected at a nearby location outside the 6-mile segment modeled.

Other Considerations. Data limitations—such as number/location of collection points or level of data aggregation—can limit a modeler's ability to accurately recreate a network, thereby influencing the accuracy of a model. There existed minor data limitations for I-91 (such as traffic composition) that may have reduced the accuracy of the VISSIM network created.

Work Zone

The work zone evaluated was a part of the Massachusetts Department of Transportation's I-91 Viaduct Rehabilitation Project near Springfield, MA, which replaces the existing deck of the bridge (*29*). This analysis takes place during Stage 1B of the project (December 2015 to November/December 2016). During Stage 1B, the inside lane (3rd lane) of Interstate 91 (I-91) northbound (NB) was closed, starting three quarters of a mile north of Exit 6 until the Interstate 291 (I-291) eastbound (EB) on-ramp (see Figure 4).

VISSIM DLL Interface

The FHWA Work Zone Driver Model DLL v1.0 was interfaced with VISSIM using the existing DLL interface in VISSIM. Link numbers for the VISSIM (FHWA) network were changed to designate the links for advanced warning, taper zone, and work zone.

It is important to note that when a car-following DLL is interfaced with VISSIM, it overrides all acceleration/deceleration functions. A DLL cannot be created to override just car-following, nor can it; a DLL cannot be created to return VISSIM's acceleration/deceleration for other behaviors or scenarios (such as slowing/stopping at a red traffic signal, slowing/yielding to other vehicles in a conflict area, or slowing to accommodate lane changing for route choice or cooperative lane changes). Ergo, if a DLL does not include acceleration/decerleration functions for these interactions, then vehicles will not slow for these scenarios or elements.

The FHWA Work Zone Driver Model v1.0 includes acceleration/deceleration algorithms for car-following only and does not include acceleration/deceleration algorithms for other behaviors. Future versions of the FHWA Work Zone Driver Model DLL will include algorithms for elements (such as conflict zone) and behaviors (such as lane changes for route choice) commonly

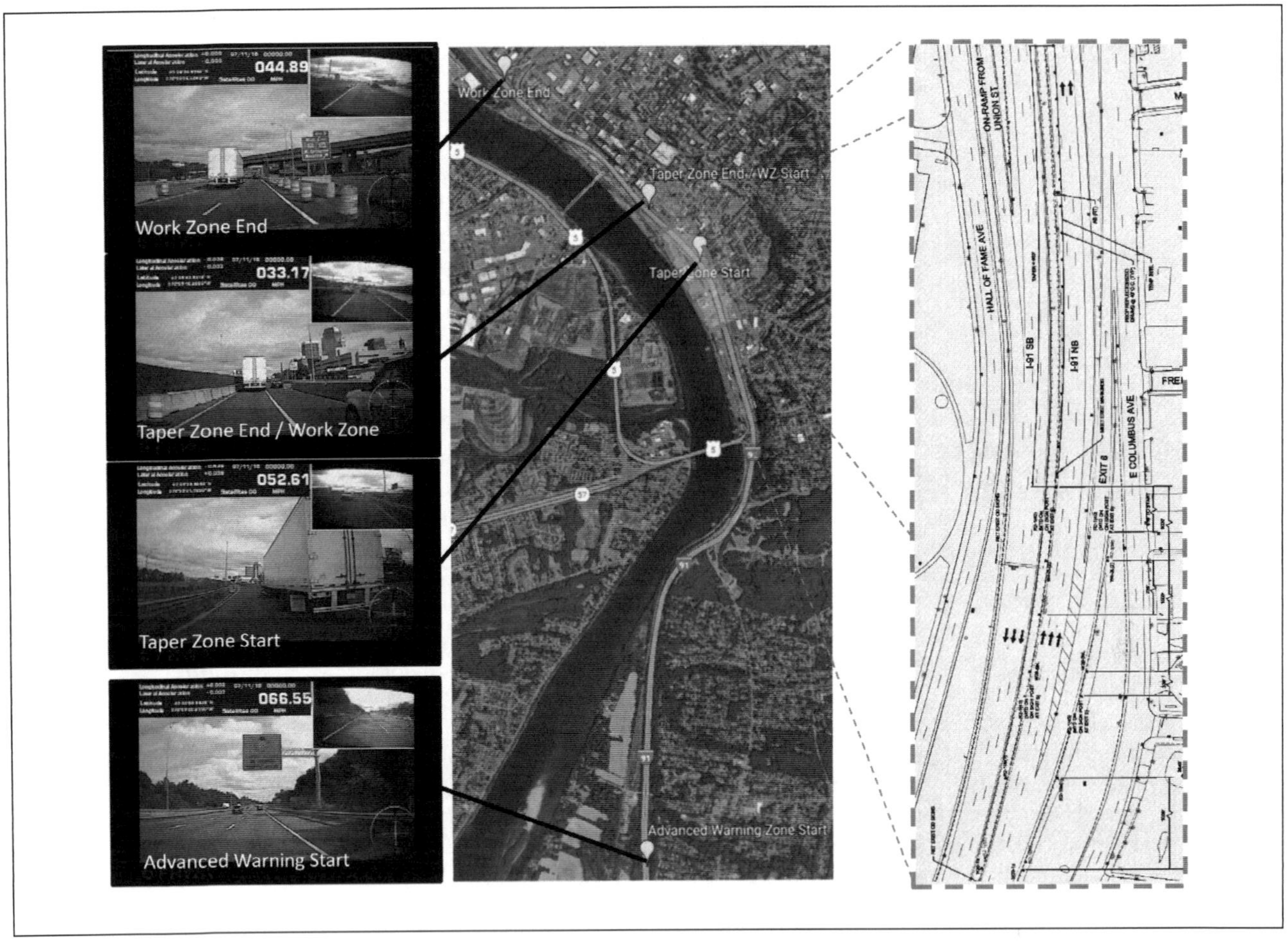

Figure 4. The I-91 NB closure, as seen from the IRV (left), satellite (middle), and MassDOT AutoCAD drawings of the lane closure (right).

found near work zones. However, when using version 1.0, network adjustments should be made to avoid error.

Taper Zone Network Adjustments. Without acceleration/deceleration algorithms for conflict zones, vehicles traversing the closing lane will not slow or yield to vehicles that have the right-of-way; and without algorithms for cooperative merging, all vehicles merge at full speed. In earlier tests of the FHWA DLL, this created unrealistic merging behaviors at the taper zone. This problem is corrected by adding reduced speed zones to the closing lane, adjacent lanes, and the links approaching them (1,000 feet before lane closure). This solution helps simulate more realistic lane-change speeds at the start of the taper zone.

Unfortunately, this solution causes all vehicles to slow uniformly—not just the vehicles yielding during a merge. This may reduce variance in travel time and queue lengths predicted by the FHWA model near the taper zone.

Static Route Network Adjustments. Static routes were used in the I-91 network. In VISSIM, if a vehicle enters a link that is not along its route, VISSIM deletes the vehicle.

Without deceleration algorithms to accommodate lane changing for route choice, some DLL vehicles failed to slow and change lanes to stay on their route. They missed a required lane change, entered a link that was not on their route, and were deleted. This created problems in earlier tests, as over 1,500 vehicles were deleted from the network (more than 1/6 the entire volume), drastically reducing volume and placing traffic in near free-flow. To correct the problem, vehicle look-back distances were adjusted on each link so that vehicles merge early to stay on their static route. Additionally, on links with disappearing vehicles, vehicle inputs were increased by the

Table 2. Summary Table: Travel Time through each Work Zone Segment: Field vs. VISSIM (Wiedemann) vs. VISSIM (FHWA)

		Average travel time [min:s]			
		Advanced warning zone	Taper + buffer zone	Work zone	Total
Field	Mean	4:01	0:47	1:29	6:18
	SD	1:02	0:07	0:30	1:09
VISSIM (Wiedemann)	Mean	4:27	0:22	1:44	6:33
	SD	0:19	0:00	0:02	0:20
VISSIM (FHWA)	Mean	4:18	0:36	2:07	7:02
	SD	0:01	0:00	0:01	0:03

Note: SD = standard deviation.

number of disappearing vehicles so that the total network volume remained unchanged.

Evaluation: Methodology and Results

The performance of the FHWA Work Zone Driver Model v1.0 was compared with field observations and with the performance of the microsimulation tool's native car-following algorithms (i.e., VISSIM's Wiedemann 99). The VISSIM models (Wiedemann and FHWA) were each run 10 times, generating results. Results were compared with field data collected from seven runs of the FHWA IRV, traversing the I-91 northbound network in July 2016.

Three performance metrics were selected: travel time through the work zone segments (i.e., advanced warning, taper zone, and work zone); travel speeds through the work zone segments (i.e., advanced warning, taper zone, and work zone); and the location of the back of queue.

Performance metrics were selected based on literature review findings, to align with metrics that state and local agencies are interested in, and to ensure that the metrics have been used to compare model performance in similar studies (thereby ensuring scientific validity of the approach).

Performance Measure 1: Travel Time through the Work Zone Segments

Travel time through the work zone segments was selected as a performance metric because: (1) DOTs reported using travel times to estimate work zone delay; (2) the travel time is directly affected by acceleration/deceleration predicted in car-following models and is therefore a useful metric for validating model performance. Travel times were collected for the advanced warning zone, the taper and buffer zone, and the work zone.

Field travel times were collected using the floating car technique in the IRV. Travel times through the advanced warning zone, taper zone, and work zone were recorded for each run. The I-91 NB network was recreated in VISSIM 7.0. The simulation was run 10 times using VISSIM's Wiedemann 99 model and the FHWA Work Zone Driver Model, and travel times through the work zone segments were recorded. The average travel time through each segment was compared (Table 2). Box-and-whisker plots were created to compare the ranges of travel times predicted and observed (Figure 5).

Travel times predicted by both models fell within (or near) the range of field-observed travel times for all three work zone segments; however, neither model reproduced the broad range of travel times observed in the field.

Performance Measure 2: Travel Speed through the Work Zone

The second performance metric, travel speed through the work zone segments, was selected because: (1) DOTs reported using travel speeds to estimate network performance; (2) travel speeds have been used to validate simulation models in prior studies, therefore this is a valid performance metric.

The range of travel speeds through each work zone segment were recorded and compared for the VISSIM simulations and the Field. Travel speeds at specific locations (such as the start of the taper zone) were also recorded and compared for the VISSIM simulations and the Field.

Travel speeds for the field data were collected by recording the range of speeds observed by the IRV as it traversed each work zone segment (i.e., the advanced warning zone, taper zone, and work zone) along I-91 NB in Springfield. The range of speeds was reported on a "per run" basis, and the average low and average high speeds across all runs was calculated. The IRV's instantaneous speed at the beginning and end of each work zone segment was also recorded.

In VISSIM, the average speed of vehicles traversing each work zone segment and the average instantaneous speed of vehicles at the beginning and end of each work

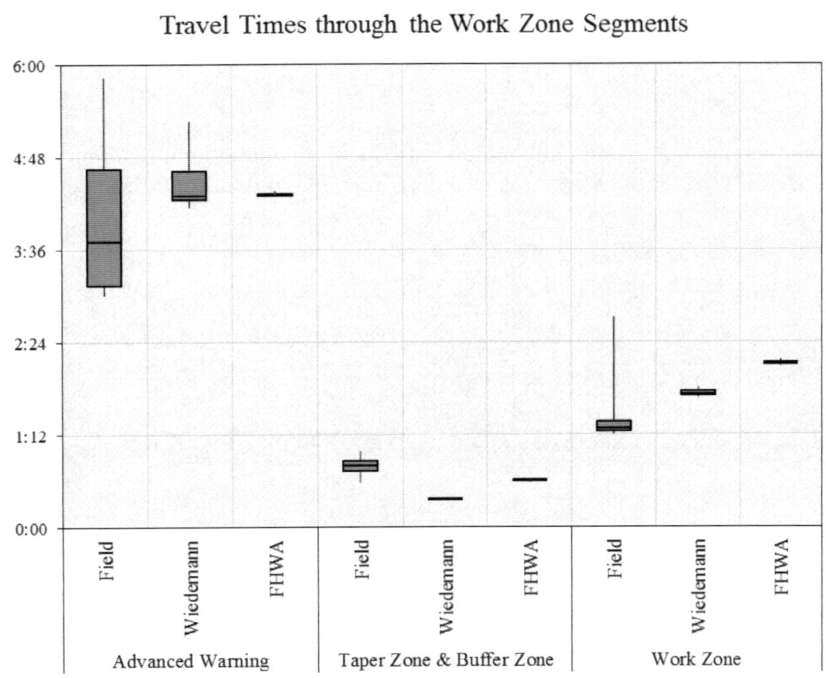

Figure 5. Comparison of travel times through all work zone segments—Field vs. VISSIM (Wiedemann) vs. VISSIM (FHWA).

Table 3. Summary Table: Travel Speeds Observed along each Work Zone Segment: Field vs. VISSIM (Wiedemann) vs. VISSIM (FHWA)

		Travel speeds [mph]		
		Advanced warning zone	Taper + buffer zone	Work zone
Field	Mean	45.8	27.1	39.4
	Avg. Low	20.9	14.4	28.4
	Avg. High	70.7	39.7	50.4
VISSIM (Wiedemann)	Mean	37.6	17.5	49.6
	Avg. Low	0	0	0
	Avg. High	84.4	84.9	83.9
VISSIM (FHWA)	Mean	47.8	36.5	40.1
	Avg. Low	0	0	12.8
	Avg. High	81.8	74.1	76.1

Note: Avg. = average.

Table 4. Summary Table: Travel Speeds at the Beginning and End of each Work Zone Segment: Field vs. VISSIM (Wiedemann) vs. VISSIM (FHWA)

		Travel speeds [mph]			
		Entry point, advanced warning zone	Entry point, taper zone	Entry point, work zone	Exit point, work zone
Field	Mean	67.3	26.1	37.1	50.6
	SD	3.7	14.9	1.9	2.9
VISSIM (Wiedemann)	Mean	65.0	54.8	58.3	54.4
	SD	0	1.8	1.1	1.4
VISSIM (FHWA)	Mean	65.0	36.1	29.8	56.0
	SD	0	0.7	0.4	0

Note: SD = standard deviation.

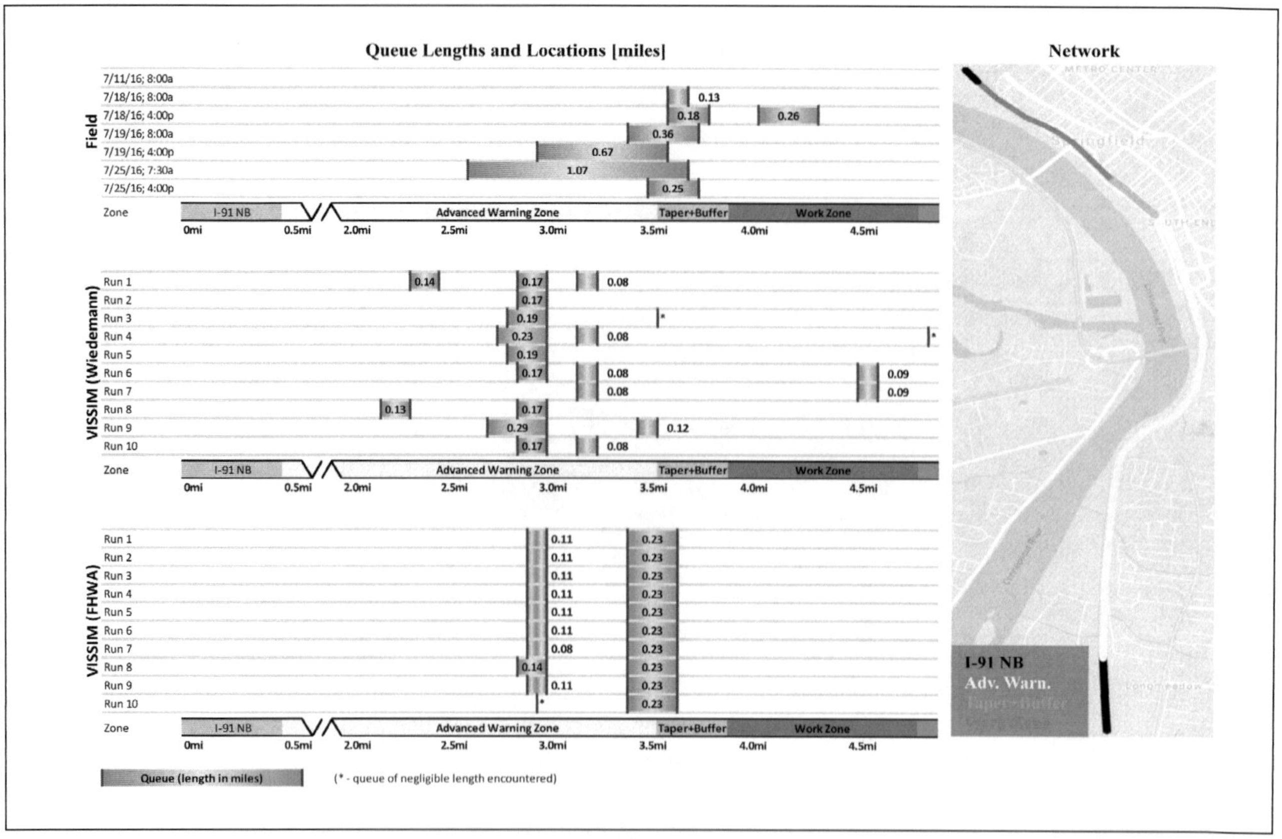

Figure 6. Queue lengths and locations per run: Field vs. VISSIM (Wiedemann) vs. VISSIM (FHWA).

zone segment was recorded for the Wiedemann and FHWA models.

Summary results were compiled and compared—Field vs. VISSIM (Wiedemann) vs. VISSIM (FHWA)—for average speed per work zone segment (Table 3) and average instantaneous speed measured at the beginning and end of each work zone segment (Table 4).

The FHWA model more accurately predicted average travel speeds; however, both models failed to recreate the variance in travel speeds observed, especially at the taper zone entry point.

Performance Measure 3: Back of Queue

The third performance metric, the back-of-queue locations, was selected because: (1) state and local DOTs report using back-of-queue predictions to determine how and where to deploy queue warning (therefore accurate back-of-queue predictions would be important to DOT agencies); (2) queue length has been used in peer-reviewed publications as a performance metric in the evaluation of macroscopic work zone models—such as QUEWZ, QuickZone, and FREEVAL—and in microsimulation to estimate impacts for work zone planning efforts (*9*).

For the purposes of this study, a queue was defined as an instance in which vehicles were forced to reduce speed to 25 mph or less due to local congestion, and the queue length was the distance between the first and last queued vehicle.

To estimate the presence and location of queues in the field, IRV GPS data and video recordings were reviewed. Seven queues were observed; GPS coordinates for the back and front of each queue were collected and mapped.

The same method was used to determine presence and location of queues in VISSIM, for the FHWA model and Wiedemann model. Simulations were stopped at 45 minutes. If a platoon of vehicles was slowed to 25 mph or less, the coordinates for the platoon were recorded, and the back and front of each queue were plotted. A comparison between field-observed and VISSIM-simulated queues is shown in Figure 6.

Field data showed queues forming in the taper zone. The queues predicted by Weidemann were in the wrong location (they formed in the advanced warning zone); however, it was able to recreate some of the variations in queue lengths observed in the field. FHWA predicted queues that started in the right location; however,

FHWA did not predict the variance in queue lengths observed in the field data.

Both Wiedemann and FHWA consistently predicted an upstream queue in advanced warning that was not observed in field data. The queue location corresponded to the US5 on-ramp and was likely caused by vehicles entering I-91.

Summary and Conclusion

Performance

The FHWA Work Zone Driver Model DLL v1.0 performance was acceptable. The average total travel time predicted by FHWA was less accurate than Wiedemann (Table 2); however, travel time per segment was more accurate for both the advanced warning and taper zones. Travel speeds per segment (Table 3) for FHWA were more accurate accross all segments, with average, low, and high speeds closer to those observed in the field. The average instantaneous speed at critical points (Table 4) was also closer to field obsservations for all segments. Queue locations predicted by FHWA were consistent with those observed in the field, with most queues beginning in the taper zone and extending into advanced warning; these queues were not observed in the Wiedemann model.

Both Wiedemann and FHWA failed to reproduce the per-run variance observed in the field data. For FHWA, this could be caused by the missing acceleration/deceleration algorithms, since decelerations from chance encounters at conflict and merge zones often produce shock waves in near-congested or "at capacity" conditions. This could also indicate a need for dedicated lane-changing algorithms, as lane changes sometimes result in disruptions to traffic flow.

Next Steps

Results from this case study demonstrate that both Weidemann and the FHWA Work Zone Driver Model could predict some work zone operational impacts; however, there are opportunities for improvement:

Network Volumes. Daily traffic volume fluctuations were unaccounted for in these models, potentially creating the field variations that neither model could reproduce. Adjusting input volumes "per run" to match volumes of specific days may improve model variance.

Model Improvements (FHWA Work Zone Driver Model DLL v2.0). The FHWA Work Zone Driver Model DLL v1.0 performed well, but there are opportunities to improve version 2.0 of the DLL.

Although network adjustments helped correct for the missing algorithms, the DLL should be updated to include acceleration/deceleration algorithms for behaviors other than car-following in VISSIM. These include deceleration algorithms for:

- Priority rules
- Conflict areas
- Lane change/gap acceptance
- Route choice (slowing to change lanes)
- Hook following

This may help recreate variances observed in the field and could make the model easier to use for practitioners.

The DLL could also be upgraded with a work zone lane-changing model, as work zone lane changing is often a specialized behavior.

Optimize Utilization. Since Weidemann includes deceleration algorithms that FHWA is missing (such as cooperative merging and lane change), using the FHWA model and Weidemann model in mixed traffic could optimize results. Weidemann vehicles would introduce decelerations at merge locations, and the FHWA model could improve travel time predictions. Additional model runs and new case studies are required to find an optimal solution for using both models.

Author Contributions

The authors confirm contributing to the paper as follows: study conception and design: A. Berthaume, C.L. Melson; data collection: A. Berthaume, L. Jackson, I. Berg, B. O'Donnell; analysis and interpretation of results: A. Berthaume, L. Jackson, I. Berg, B. O'Donnell; draft manuscript preparation: A. Berthaume, L. Jackson, I. Berg. All authors reviewed the results and approved the final version of the manuscript.

References

1. Federal Highway Administration. *Highway Statistics 2015*. 2015. https://www.fhwa.dot.gov/policyinformation/statistics/2015/. Accessed July 1, 2017.
2. Federal Highway Administration. *FHWA Work Zone Facts and Statistics*. https://ops.fhwa.dot.gov/wz/resources/facts_stats.htm. Accessed July 1, 2017.
3. Tang, C., and S. I.-J. Chien. Optimization of Work Zone Schedule Considering Time-Varying Traffic Diversion. Presented at 89th Annual Meeting of the Transportation Research Board, Washington, D.C., 2010.
4. Schrank, D., T. Lomax, and S. Turner. *2010 Urban Mobility Report*. Texas A&M Transportation Institute, 2010.
5. U.S. Department of Transportation. *Work Zone Safety and Mobility Final Rule*. https://ops.fhwa.dot.gov/wz/docs/wz_final_rule.pdf. Accessed July 1, 2017.

6. Collura, J., K. Heaslip, M. Knodler, D. Ni, W. C. Louisell, A. Berthaume, R. Khanta, K. Moriarty, and F. Wu. *Evaluation and Implementation of Traffic Simulation Models for Work Zones*. Publication NETC 05-8. New England Transportation Consortium, Burlington, VT, 2010.
7. Schnell, T., J. Mohror, and F. Aktan. Evaluation of Traffic Flow Analysis Tools Applied to Work Zones Based on Flow Data Collected in the Field. *Transportation Research Record: Journal of the Transportation Research Board*, 2002. 1811: 57–66.
8. Hardy, M., and K. Wunderlich. *Traffic Analysis Toolbox Volume IX: Work Zone Modeling and Simulation - A Guide for Analysts*. Publication FHWA-HOP-09-001. Federal Highway Administration, Washington, D.C., 2009.
9. Nikolic, G., M. Chan, and R. Pringle. Planning and Implementing a Full Closure of the Highway 401 Express Lanes with the Help of Micro-Simulation. Presented at the Annual Conference of the Transportation Association of Canada, Calgary, Canada, 2005.
10. Hallmark, S., J. Turner, and C. Albrecht. *Synthesis of Work-Zone Performance Measures*. Publication 12–436. Iowa State University Institute for Transportation, Ames, IA, 2013.
11. Michigan Department of Transportation. *MDOT Work Zone Safety and Mobility Manual*, January 2010. https://www.michigan.gov/documents/mdot/MDOT_WorkZoneSafetyAndMobilityManual_233891_7.pdf.
12. Griffith, A., and M. Lynde. *Assessing Public Inconvenience in Highway Work Zones*. Publication FHWA-OR-RD-02-20. Oregon Department of Transportation, Salem, OR, 2002.
13. Ullmann, G. L., T. J. Lomax, and T. Scriba. *A Primer on Work Zone Safety and Mobility Performance Measurement*. Publication FHWA-HOP-11-033. Texas Transportation Institute, College Station, TX, 2011.
14. Heaslip, K., A. Kondyli, D. Arguea, L. Elefteriadou, and F. Sullivan. Estimation of Freeway Work Zone Capacity Through Simulation and Field Data. *Transportation Research Record: Journal of the Transportation Research Board*, 2009. 2130: 16–24.
15. Yeom, C., A. Hajbabaie, B. J. Schroeder, C. Vaughan, X. Xuan, and N. M. Rouphail. Innovative Work Zone Capacity Models from Nationwide Field and Archival Sources. *Transportation Research Record: Journal of the Transportation Research Board*, 2015. 2485: 51–60.
16. Pringle, R., and G. Nikolic. Getting Simulation "Over the Hump" as an Operational Analysis Tool. Presented at the Annual Conference of the Transportation Association of Canada, Charlottetown, 2015.
17. Washington State Department of Transportation. *Protocol for VISSIM Simulation*, September 2014. http://www.wsdot.wa.gov/Design/Traffic/Analysis/VISSIMProtocol.htm.
18. Yeom, C., N. M. Rouphail, W. Rasdorf, and B. J. Schroeder. Simulation Guidance for Freeway Lane Closure Capacity Calibration. Presented at 95th Annual Meeting of the Transportation Research Board, Washington, D.C., 2016.
19. Chatterjee, I., P. Edara, S. Menneni, and C. Sun. Replication of Work Zone Capacity Values in a Simulation Model. *Transportation Research Record: Journal of the Transportation Research Board*, 2009. 2130: 138–148.
20. Hollander, Y., and R. Liu. The Principles of Calibrating Traffic Microsimulation Models. *Transportation*, Vol. 35, No. 3, 2008, pp. 347–362. https://doi.org/10.1007/s11116-007-9156-2.
21. Brockfeld, E., R. Kühne, and P. Wagner. Calibration and Validation of Microscopic Traffic Flow Models. *Transportation Research Record: Journal of the Transportation Research Board*, 2004. 1876: 62–70.
22. Punzo, V., and F. Simonelli. Analysis and Comparison of Microscopic Traffic Flow Models with Real Traffic Microscopic Data. *Transportation Research Record: Journal of the Transportation Research Board*, 2005. 1934: 53–63.
23. Wu, J., M. Brackstone, and M. McDonald. The Validation of a Microscopic Simulation Model: A Methodological Case Study. *Transportation Research Part C: Emerging Technologies*, Vol. 11, No. 6, 2003, pp. 463–479. https://doi.org/10.1016/j.trc.2003.05.001.
24. Lochrane, T. *A New Multidimensional Psycho-Physical Framework for Modeling Car-Following in a Freeway Work Zone*. PhD dissertation. University of Central Florida, Orlando, FL, 2014.
25. Berthaume, A. L. *Microscopic Modeling of Driver Behavior Based on Modifying Field Theory for Work Zone Application*. PhD dissertation. University of Massachusetts, Amherst, MA, 2015.
26. Lochrane, T. W. P., H. Al-Deek, D. J. Dailey, and J. Bared. Using a Living Laboratory to Support Transportation Research for a Freeway Work Zone. *Journal of Transportation Engineering*, Vol. 140, No. 7, 2014, pp. 0401–4024. https://doi.org/10.1061/(ASCE)TE.1943-5436.0000674.
27. Federal Highway Administration. FHWA Delivers Instrumented Vehicle to the Volpe Center. *FHWA R&T Now*, July/August 2016. https://www.fhwa.dot.gov/publications/rtnow/16jul_aug_rtnow.pdf.
28. Massachusetts Department of Transportation. *Traffic Volume Counts*. https://www.massdot.state.ma.us/highway/TrafficVolumeCounts.aspx. Accessed July 1, 2017.
29. Massachusetts Department of Transportation. *Interstate 91 Viaduct Rehabilitation Project*. http://www.massdot.state.ma.us/i91viaductrehab/Home.aspx. Accessed July 1, 2017.

The Standing Committee on Work Zone Traffic Control (AHB55) peer-reviewed this paper (18-05790).

Article

TRR
JOURNAL OF THE TRANSPORTATION RESEARCH BOARD

Transportation Research Record
2018, Vol. 2672(16) 57–67

DOI: 10.1177/0361198118796736
journals.sagepub.com/home/trr
SAGE

Closed Course Performance Testing of a Work Zone Intrusion Alarm System

LuAnn Theiss[1], Tomas Lindheimer[1], and Gerald L. Ullman[1]

Abstract

Oldcastle Materials, Inc. and ARTIS, LLC joined forces to develop an innovative work zone intrusion detection and alarm system. The alarm system flashes light-emitting diodes and emits an audible alarm if an intrusion is detected. Texas A&M Transportation Institute performed a closed course study to assess performance of the alarm system under conditions that may occur in real work zones. This testing was intended to verify that the alarm system does produce the proper alert when conditions warrant it and that alerts were not activated when conditions did not warrant them (i.e., that the system does not produce false alarms). Performance of the worker body alarms (designed to be worn by field personnel and activated when an intrusion threat is detected) was also evaluated. During testing, these devices were positioned at various locations to assess the ability of the system to correctly determine the location of the devices and their position relative to the intrusion threat. The alarm system was tested under two operating modes: lane closures and flagging operations. Several approach vehicle trajectories were used in both operating modes. The main alarm achieved a 100% success rate, whereas the worker body alarms achieved a 97% success rate. Unsuccessful events for the worker body alarms were attributable to the test protocol exceeding the effective communication range of the intrusion detection system. Ongoing improvements in the device are expected to significantly increase this communication range. All other device locations, scenarios, and trajectories tested achieved a 100% success rate.

Although work zone intrusion crashes only comprise a small portion (7% to 10%) of all crashes that occur in work zones, they are a major concern to both transportation agencies and contractors (*1, 2*). Efforts to develop an effective means of detecting work zone intrusions and warning workers with an audible alarm have existed since the late 1980s (*3*). Early systems utilized pneumatic tubes or infrared beams placed along the edge of the work space, and activated an alarm for workers if a vehicle crossed the tube or broke the beam (*4*). Another design attached the alarm to channelizing devices that activated if the device was knocked over (*5*). However, these systems all suffered from frequent false alarms. More recently, newer directional audible system (DAS) technology has been developed and tested in a mobile work zone field deployment (*6*). One study evaluated the accuracy of an alarm with DAS technology and found that the system performed poorly in horizontal curve situations (*7*).

Oldcastle Materials and ARTIS, LLC joined forces to develop an innovative work zone intrusion detection and alarm system. This new alarm system utilizes a target threat detection and tracking device (the main alarm) that is capable of logically assessing approaching vehicle speed, location, and possible trajectory. If the trajectory is computed to result in the intrusion of the vehicle into the work zone, flashing light-emitting diode (LED) warning lights and an audible alarm are activated to alert the motorist and also notify workers of the intrusion. In addition to warning work crews of an intrusion, the expectation is that the alarm system will reduce intrusion consequences by catching the attention of the targeted approaching motorist with the activation of the flashing LED lights and the audible alarm. The alarm system also includes worker body alarms that can be worn as part of a worker's personal protective equipment (PPE). The worker body alarms can be worn in a pocket, on a vest, or mounted on a hardhat. When the main alarm system is triggered and the worker is positioned within the potential trajectory of the intrusion threat vehicle, the worker body alarm produces both vibratory (tactile) and audible alerts. Together, it is hoped that these countermeasures will deter vehicle intrusions into the work

[1]Texas A&M Transportation Institute, College Station, TX

Corresponding Author:
Address correspondence to LuAnn Theiss: l-theiss@tti.tamu.edu

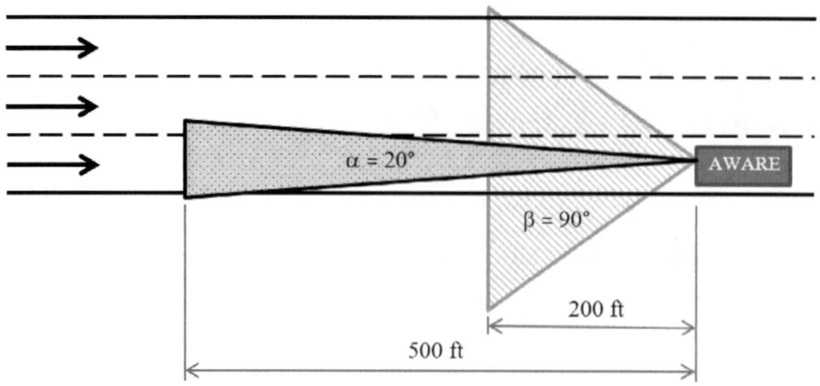

Figure 1. Threat detection regions.

space, alert workers about the event, and increase everyone's chance of escaping unharmed.

Description of the Alarm System

When the alarm system is deployed in a roadway environment (i.e., a work zone), threats are detected in two flat, fan-shaped regions, as shown in Figure 1.

The long-range region extends approximately 500 ft upstream of the main alarm system. The width of this region expands 10° on either side of the centerline (for a total detection angle $\alpha = 20°$). The short-range region extends approximately 200 ft upstream of the alarm system and expands 45° on either side of the centerline (for a total detection angle $\beta = 90°$).

When a vehicle enters these protected areas, its speed and heading are detected by the alarm system, which uses internal calculations to determine the appropriate response. The calculations are based primarily on the stopping sight distance (SSD) equation from the American Association of State Highway and Transportation Officials (AASHTO) *A Policy on Geometric Design of Highways and Streets*, commonly known as the AASHTO Green Book (*8*):

$$SSD = 1.47Vt + 1.075\frac{V^2}{a}$$

where

t = the perception-reaction time of 2.5 s;
a = the deceleration rate of 11.2 ft/s^2; and
V = the approach vehicle closing speed in mph.

The equation assumes that sufficient tire–pavement friction exists to create the stated deceleration rate. In addition, the assumed values for perception–reaction time and deceleration rate are characteristic of 90% of the driving population. Thus, for a vehicle approaching at a speed of 45 mph, the calculated AASHTO SSD would be 360 ft; for a vehicle approaching at 60 mph, the AASHTO SSD would be 566 ft.

The alarm system also relies on an alternate (more conservative) calculation of SSD assuming that t = 4.5 s to activate some of its alerts and alarms. By substituting this in the equation in lieu of 2.5 s, the outcome is longer SSD values. For example, for a vehicle approaching at a speed of 45 mph, the conservative calculated SSD would be 492 ft; for a vehicle approaching at 60 mph, the conservative SSD would be 742 ft. As shown in Figure 2, the conservative SSD is longer than the long-range threat detection capabilities of the alarm system once approach speeds exceed 45 mph, and the AASHTO SSD is longer than the long-range region once approach speeds exceed 55 mph. This means that the detection range of the system governs activation of its alarms and alerts for system testing purposes.

If the trajectory of the vehicle is computed to intrude into the work space/protected area (or to be exceeding a reasonable speed approaching the work zone), the alarm system is activated to alert the motorist and also notify workers of the intrusion threat. If the conservative SSD threshold corresponding to t = 4.5 s is exceeded, the flashing LED warning lights activate. If the SSD threshold corresponding to t = 2.5 s is exceeded, then both the flashing lights and an audible alarm are activated.

Researchers at Texas A&M Transportation Institute (TTI) conducted system performance testing of the intrusion alarm system, including the worker body alarms. The objective of this testing was intended to verify that the alarm system does produce the proper alert when conditions warrant it (i.e., lights activate and do so when the approaching vehicle is at the appropriate distance based on the threat detection and SSD algorithms). In addition, the testing was also intended to verify that alerts were not activated when conditions did not

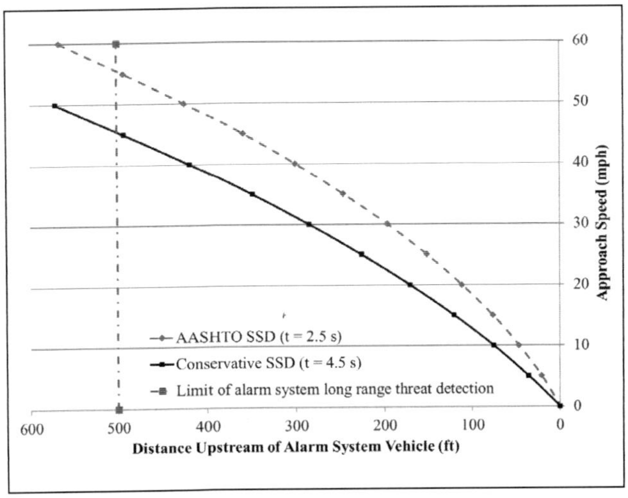

Figure 2. AASHTO SSD and conservative SSD values used by the alarm system.
Note: SSD = stopping sight distance; AASHTO SSD= Association of State Highway and Transportation Officials stopping sight distance.

warrant them (i.e., that the system does not produce false alarms).

Methodology

The testing was performed at the Texas A&M University RELLIS Campus in August of 2016 and consisted of TTI researchers driving instrumented approach vehicles toward the main alarm system under various conditions to determine whether the system responded as expected. The alarm system developers and technicians were also on site. They installed, armed, and otherwise operated the alarm system. TTI staff installed and operated separate data collection equipment to observe system responses. Subsequent data analyses were performed to assess actual distances at which alarms were activated to further validate the precision of the alarm system algorithms.

Data Collection Equipment

During testing, the researchers used two different approach vehicles (a passenger car and a pickup truck), which allowed the researchers to verify that the alarm system could detect vehicles of different sizes. The approach vehicles were instrumented with a Trimble global positing system (GPS) using real time kinematic (RTK) satellite navigation to continuously track the approach vehicle position via corrected GPS data. The RTK base station was located approximately 3,200 ft (less than1 km) from the test site, which reportedly provided positional accuracy of the instrumented vehicle within 0.4 in. The GPS data were captured by an in-vehicle laptop. A dash-mounted video camera was also linked to the in-vehicle laptop for continuous capture of the forward scene view from the approach vehicle. As a backup, a separate time-synced video camera also independently recorded the forward scene data. During portions of the testing, worker body alarms were also evaluated for proper response when the main alarm was triggered. The researchers used tripod-mounted video cameras to record the worker body alarm responses (i.e., alarm sound and vibration) using time-synchronized videos that could then be matched to the time stamps of the in-vehicle data.

Lane Closure in Tangent Alignment

During this portion of the testing, the alarm system was mounted on the roof of a work vehicle (another pickup truck) which was parked in a staged lane closure. Figure 3 shows the approach path trajectories used for this portion of the testing.

In Trajectory A, TTI staff drove the instrumented vehicle toward the alarm system beginning two lanes left of the lane in which the alarm vehicle was located. Once the instrumented vehicle was within 200 ft of the alarm vehicle (i.e., within the short-range detection region of the system), the TTI driver made an abrupt lane change into the lane adjacent to the alarm vehicle, and then continued past the alarm vehicle in that adjacent lane. The expectation was that the alarm system would not produce an alert (i.e., would not activate the lights or audible alarm).

In Trajectory B, the instrumented vehicle began in the lane adjacent to the lane in which the alarm system vehicle was located. The instrumented vehicle passed by the alarm system vehicle in the adjacent lane. Similar to Trajectory A, the alarm system was not expected to produce an alert for this trajectory.

In Trajectory C, the instrumented vehicle began in the left lane adjacent to the lane in which the alarm vehicle was located. At a location somewhere between 360 ft and 200 ft upstream of the alarm vehicle, the instrumented vehicle crossed over into the lane adjacent to the alarm vehicle on the right side. The stars indicate the worker body alarm locations for this series of tests. The expectation was that this trajectory would produce an alert that would trigger the lights and audible alarm. The worker body alarms would activate only if the threat was detected upstream of the worker body alarm locations (i.e., the worker body alarms located at 0 ft and potentially 300 ft upstream would produce an alert, but the worker body alarm at 400 ft would not).

In Trajectory D, the instrumented vehicle began in the same lane in which the alarm system vehicle was located and traveled toward the alarm system at a speed of 15 mph. Within 100 ft of the alarm system vehicle, the

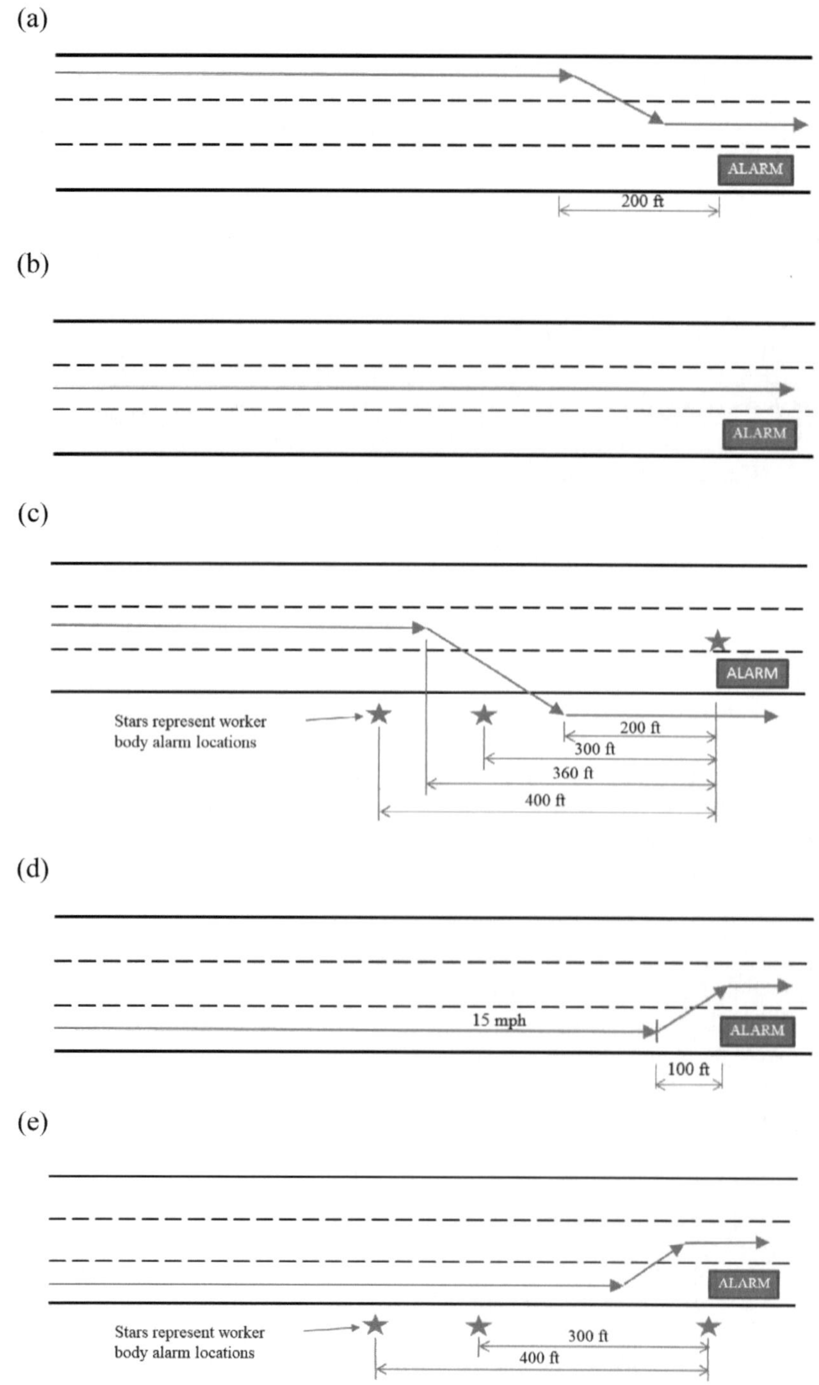

Figure 3. Trajectories for lane closure in tangent alignment testing: (a) Trajectory A: lane change within 200 ft of alarm; (b) Trajectory B: pass by in adjacent lane; (c) Trajectory C: crossing within intrusion detection region; (d) Trajectory D: approach speed below threshold speed; (e) Trajectory E: SSD penetration.

instrumented vehicle came to a full stop, and then passed the alarm vehicle on the left at a speed no greater than 15 mph, similar to how a construction vehicle might enter and pass by a work crew. For this trajectory, the expectation was that the alarm system would not produce an alert. Once an appropriate system response was verified using the instrumented passenger car, the sponsor agreed that the pickup truck need not also be tested under this trajectory.

In Trajectory E, the instrumented vehicle penetrated the SSD threshold for a prescribed approach speed. When the approach speed was 45 mph, the AASHTO SSD threshold was 360 ft and the conservative SSD was 492 ft. When the approach speed was 60 mph, the AASHTO SSD threshold was 566 ft and the conservative SSD was 742 ft. Again, the stars indicate the worker body alarm locations for this series of tests. The expectation was that this trajectory would produce an alert that would trigger the lights, audible alarm, and worker body alarm components of the system. For both approach speeds, the lights and worker body alarm devices should be activated upon penetration of the long-range threat detection region as the conservative SSD in both cases was equal to or greater than the 500 ft long-range threat detection region.

Each of the trajectories was replicated at least three times at each speed in each instrumented approach vehicle with the alarm system vehicle parked parallel to the lane lines. The researchers also evaluated the ability of the alarm system to operate in slightly skewed deployments. This was intended to simulate a condition where the alarm system vehicle may not be perfectly aligned during deployment in a real work zone. These data were collected with the alarm system vehicle alignment skewed approximately 10° left and 10° right from parallel to the travel lane. In addition, data were collected with the alarm vehicle aligned parallel with the lanes, but with the left tires lifted approximately 4 in. off the ground to simulate the alarm system vehicle being parked on a paved shoulder (at approximately 6° of tilt).

Overall, there were 16 different test conditions for Trajectories A, B, C, and E and four different test conditions for Trajectory D. Accounting for the multiple passes using each instrumented vehicle (except for Trajectory D), a total of 204 test runs were completed during the tangent alignment testing.

Lane Closure in Curved Alignment

Curved alignment tests were conducted to determine whether the alarm system would respond properly in roadway conditions where a horizontal curve is present. A horizontal curve equivalent to a 40 mph design speed was marked on the test area pavement. This curve was believed to be sufficient to test the ability of the alarm system to properly assess whether the trajectory of the approaching vehicle would pose a threat for the work zone.

Only one instrumented approach vehicle (the passenger car) was used for the curved alignment testing. Based on observations of the tangent alignment lane closure testing, the researchers concluded that (1) the alarm system had no apparent difference in activation between the two different approach vehicles, and (2) the smaller approach vehicle would present a worst-case challenge of detection and trajectory tracking by the system. Thus, the pickup truck was not used as an instrumented approach vehicle in the curved alignment testing. In addition, only the 0° and tilted alarm vehicle alignments were evaluated during this testing (10° left and right were omitted).

The alarm system was mounted on the roof of a work vehicle (a pickup truck) that was parked in the staged lane closure. With the alarm system vehicle in position, TTI researchers drove the instrumented vehicle along a designated path at 40 mph according to four trajectories (A, B, C, and E). These trajectories were identical to those used in the tangent alignments, except they were located on right and left curves. A total of 48 test runs were made during the curved alignment testing.

Flagging Operations

For the flagging operation, the alarm system was mounted on a hand truck to replicate a freestanding "flagger cart" that was under development during this testing. The cart was positioned approximately 2 ft from the travel lane used for the instrumented vehicle approaches (to simulate where a flagger would stand). The researchers used a sledge hammer to apply a dead weight to the flagger cart foot pedal. When the pedal was pressed, the cart operated in SLOW PADDLE mode. This simulated a flagger holding the pedal down and allowing vehicles to pass by. When the sledge hammer was removed, the cart operated in STOP PADDLE mode. This simulated a flagger holding up traffic and not allowing vehicles to pass by.

Three approach runs were made in the instrumented vehicle for each trajectory. The trajectories for the flagging operation differed from those used for the lane closures. The flagging operation trajectories are shown in Figure 4.

In testing Trajectories F and G, TTI staff drove the instrumented vehicle toward the flagger cart in the approach lane at 45 mph. Once the instrumented vehicle was within 360 ft of the flagger cart, the TTI driver made a lane change into the adjacent (left) lane and completed that maneuver before reaching a point 150 ft upstream of

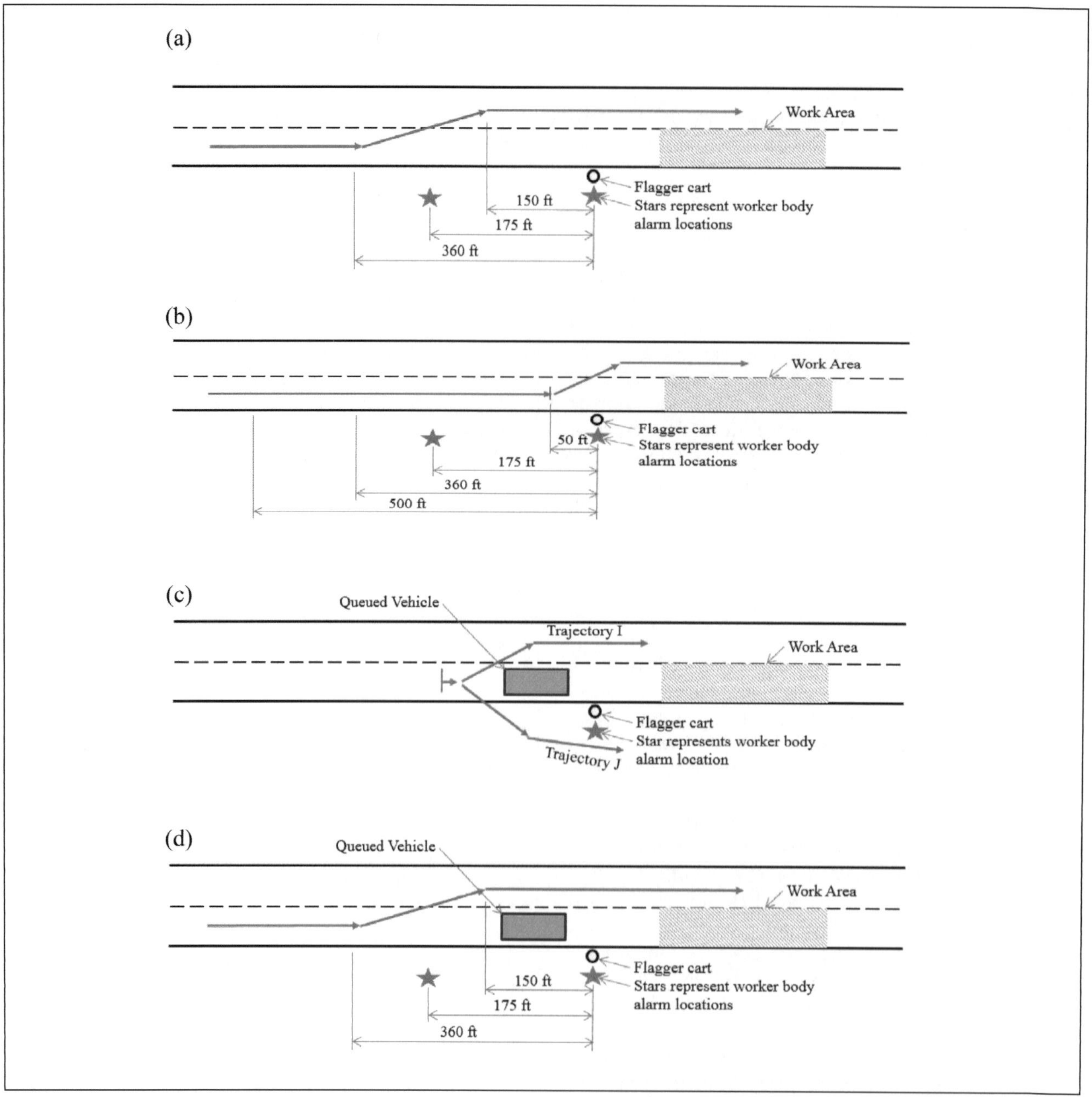

Figure 4. Trajectories for flagging operation testing: (a) Trajectories F and G: lane change with SLOW/STOP PADDLE; (b) Trajectory H: SSD threshold; (c) Trajectories I and J: passing queued vehicle on left/right; (d) Trajectory K: lane change with STOP PADDLE and queued vehicle.

the flagger cart. The driver then continued past the flagger cart in that adjacent lane at 45 mph. The flagger cart was operating in SLOW PADDLE mode during Trajectory F. This simulated a flagger waving a car around into the open lane, so the alarm system was not expected to produce an alert. During Trajectory G, the flagger cart was operating in STOP PADDLE mode. This was to simulate a vehicle disregarding the flagger instruction. The alarm system was expected to produce an alert and trigger the worker body alarm as well.

The testing of Trajectory H was intended to demonstrate that the alarm system would not respond to vehicles traveling just under the conservative SSD threshold while the flagger cart is in STOP PADDLE mode. This trajectory test would require that the driver of the instrumented vehicle continuously maintains a vehicle speed below the conservative SSD (i.e., less than 45 mph at the 500 ft mark, less than 30 mph at the 300 ft mark, etc.). This proved to be challenging in the field, so the researchers performed six runs using this trajectory, some of which exceeded the threshold and some that did not (these are described in greater detail in the Results section).

In the testing of Trajectories I and J, the flagger cart was operating in STOP PADDLE mode. A TTI fleet vehicle was parked in the approach lane approximately 20-ft upstream of the flagger cart. This was intended to simulate a condition in which a queue of waiting traffic would potentially block detection of a vehicle jumping the queue and proceeding into the work zone (i.e., ignoring the STOP PADDLE instructions).

TTI staff began the tests with the instrumented vehicle stopped in the approach lane just behind the queued vehicle. For Trajectory I, the driver accelerated and passed the queued vehicle on the left. For Trajectory J, the driver accelerated and passed the queued vehicle on the right. In both cases, the alarm system was expected to produce an alert that would trigger the lights, audible alarm, and worker body alarm components of the system.

In testing Trajectory K, the researchers repeated the Trajectory G test (with the flagger cart in STOP PADDLE mode) and added the queued vehicle (similar to Trajectories I and J). The approach speed was 45 mph. This would test whether the alarm system could still detect a high-speed lane change (representing a driver's disregard of flagger instruction to stop) around the queued vehicle. In this case, the alarm system was expected to produce an alert that would trigger the lights, audible alarm, and worker body alarm components of the system. A total of 21 test runs were made during the flagging operations testing: three runs for each of the five different trajectories (F, G, I, J, and K) and six runs for Trajectory H.

Results

Lane Closure in Tangent Alignment

For the tangent alignment trajectories, the alarm system did not activate during Trajectories A, B, and D for any of the test conditions, all of which were the appropriate responses. For Trajectories C and E, the alarm system was expected to activate the lights and audible alarm.

Where appropriate, the researchers calculated the distance between the GPS coordinates of the alarm system and those of the approach vehicle at the time of the flashing light activation. The corresponding vehicle speed at the time of the flashing light activation was noted and verified to be within 1.3 mph of the intended speed (45 mph or 60 mph).

In Trajectory C testing, the worker body alarm located adjacent to the alarm system vehicle was expected to activate, but the worker body alarm located at 300 ft and 400 ft upstream of the vehicle were not expected to activate. As stated previously, the trajectory vehicle entered the protected lane between 360 ft and 200 ft from the alarm system vehicle (typically around or closer than 300 ft) and therefore did not activate the upstream devices. Of 144 potential worker body alarm responses, 15 could not be accurately ascertained due to TTI equipment failure. However, all of the remaining responses were correct.

In Trajectory E testing, the approach vehicle made a direct penetration of the SSD in the lane where the alarm vehicle was parked. Due to road noise inside the instrumented vehicle at higher speeds coupled with being at a further distance from the alarm than in Trajectory C, the time at which the audible alarm was activated could not be determined for all test runs. Therefore, the researchers focused solely on the activation of the flashing lights and the worker body alarms for the analysis of Trajectory E data.

At 45 mph, the flashing light activation distance range was 382 to 522 ft. Statistically, there was no difference in activation distances between the car and the truck. Overall, the average activation distance was 478.8 ft. Considering that the data collection methods may have introduced some lag time due to laptop recording speed, and so forth, the researchers concluded that the alarm system met the performance requirements of a 500 ft detection range for Trajectory E at 45 mph.

At 60 mph, the flashing light activation distance range was 388 to 554 ft. Statistically, there was no difference in activation distances between the car and the truck. Overall, the average activation distance was 478 ft, essentially equal to that observed for the 45 mph tests. Once again, this demonstrates the verification of the 500 ft detection range of the system.

With regards to the worker body alarms for Trajectory E, the researchers expected all devices to activate as the trajectory vehicle entered the protected lane at the upstream end of the long-range threat detection region. At the 0 ft and 300 ft locations, the worker body alarms activated each time as expected. There were three additional instances in which TTI data collection equipment malfunctioned and the researchers could not verify that the device activated during these runs.

The worker body alarm positioned at the 400 ft location experienced mixed responses, most likely because the test protocol was exceeding the communication range limits of the system at the time of the test. At 45 mph, the worker body alarm activated during 17 of 24 runs; at

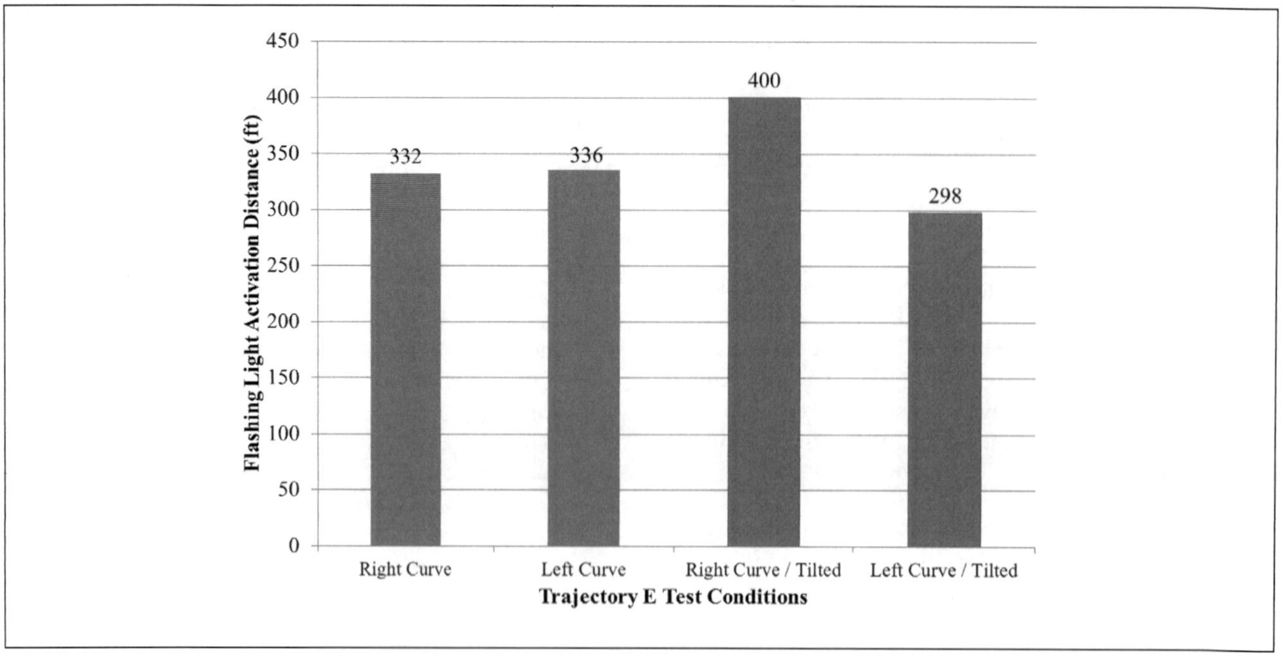

Figure 5. Curved alignment Trajectory E test results.

60 mph, the worker body alarm activated during 14 of 18 runs (with data collection equipment malfunctions not allowing verification of six runs). Work continues on the system to improve the communication range of the device, which will undoubtedly improve its performance at this farther distance.

Lane Closure on Curved Alignment

For the curved alignment tests, the alarm system did not activate during any of the A and B trajectories, which was the appropriate response. Trajectory D was not tested in the curved alignments. For Trajectories C and E, the alarm system was expected to activate. Again, the researchers calculated the distance between the GPS coordinates of the alarm system and those of the approach vehicle at the time of the alarm activation. The corresponding vehicle speed at the time of the alarm activation was noted and verified to be within 1 mph of the intended speed (40 mph).

In Trajectory C testing, the approach vehicle made a crossover maneuver and the alarm system responded with both flashing lights and sound activation as expected. The worker body alarm located adjacent to the alarm vehicle was expected to activate, and did so in all 12 runs for this trajectory. The worker body alarms located at 300 ft and 400 ft should not have activated as the crossover maneuver did not occur until the trajectory vehicle was within 300 ft of the alarm vehicle. At these locations, the proper response (no alert) was documented during all 12 runs. Thus, all responses were correct.

In Trajectory E testing, the approach vehicle made a direct penetration of the conservative SSD in the lane where the alarm vehicle was parked. As before, the researchers focused solely on the activation of the flashing lights for analysis of Trajectory E data. At 40 mph, the range of flashing light activation distance was 265 to 424 ft, measured along the curve trajectory. Figure 5 shows the results of the curved alignment testing. It appears that the direction of the curve had no impact on the alarm system's ability to detect the passenger car. Note that the conservative SSD for 40 mph is 418 ft and the AASHTO SSD is 301 ft. Thus, the system did activate prior to the AASHTO SSD value but after the conservative SSD value, on average, when the alarm system vehicles were level. Interestingly, when the alarm system vehicle was tilted, there was a noticeable difference in the activation distances. With the alarm system vehicle tilted, the activation distance in the right curve is substantially longer than in the left curve. This is likely due to the small passenger car approaching the alarm system underneath the fan-shaped threat detection region in the left curve and proceeding much closer to the alarm system before it is detected. Conversely, on the right curve, the tilted alarm system detection area appeared to activate at a distance closer to the conservative SSD than the AASHTO SSD value. Nonetheless, the alarm system did detect the instrumented vehicle and produce an alert

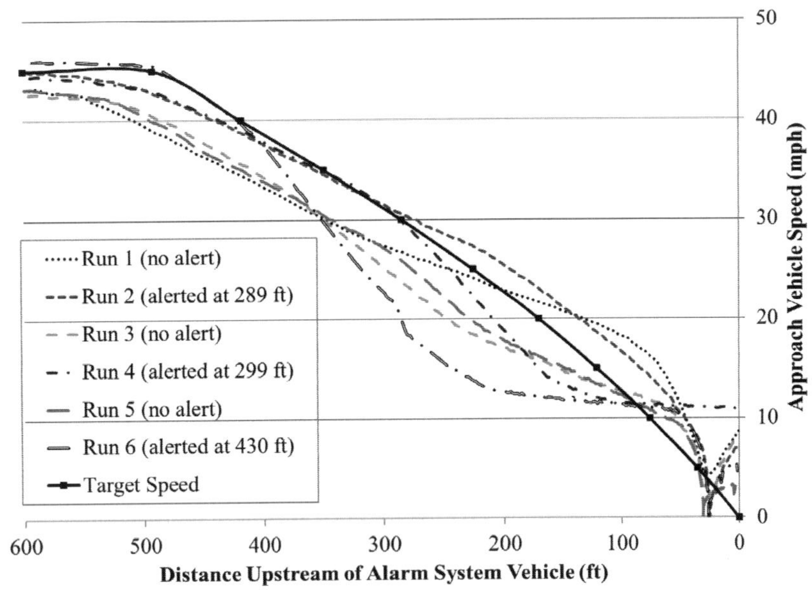

Figure 6. Test runs versus conservative SSD threshold for Trajectory H.

prior to the instrumented vehicle reaching the AASHTO SSD in all cases.

The worker body alarm located adjacent to the alarm vehicle was expected to activate, and did so in all 12 runs for Trajectory E. Because of the issues already identified during tangent tests regarding the worker body alarms when located upstream of the alarm system, they were not tested again at the 300 ft and 400 ft distances during the horizontal curve tests.

Flagging Operation

The alarm system was expected to activate during testing of Trajectories G, H, I, J, and K. The results showed that alarm activation during testing of Trajectory G occurred at an average distance of 452 ft upstream of the flagger cart. Once again, this corresponds to immediate detection upon entering the long-range (500 ft) region of the alarm system's threat detection capabilities.

The analysis of Trajectory H test results was more complex. During three of the six test runs, the alarm system lights activated, but the audible alarm did not. Further analysis of the data revealed that the instrumented vehicle speed equaled or exceeded the conservative SSD threshold at some distance upstream of the cart for some of the runs, but not others. Figure 6 illustrates those trajectories. The conservative SSD values were matched or exceeded by the instrumented vehicle at some point during test runs 2, 4, and 6. Conversely, the trajectories for runs 1, 3, and 5 remained below the conservative SSD line until speeds were below 25 mph and so correctly did not activate. Therefore, the alarm system responded appropriately in all six test runs of Trajectory H. After Trajectories F, G, and H were completed, heavy rain forced the researchers to suspend worker body alarm testing to prevent water damage to the testing equipment. Thus, worker body alarm data were collected for these 12 test runs only. The worker body alarm data recorded at the flagger station (0 ft) responded correctly in 12 of 12 runs. However, video data at the upstream worker body alarm location at 175 ft was found to be unusable due to the weather conditions that day and was not included in this analysis.

The testing of the alarm system continued with Trajectories I and J: the alarm system was able to detect the instrumented vehicle passing a queued vehicle on either side from a stopped position behind the queued vehicle. In addition, the queued vehicle did not prevent the alarm system from detecting a high-speed lane change further upstream.

Summary

Table 1 provides a summary of the alarm system performance testing results with regards to the activation of lights/audible alarm (when the audible alarm was evaluated). Based on the methodology utilized, the system achieved a 100% success rate across the range of scenarios and vehicle trajectories tested.

Table 1. Alarm System Performance Testing Results

Testing scenario	Trajectory	Number of runs performed	Number of runs with correct alarm system response	Percent correct
Tangent	A	2 vehicles × 2 speeds × 4 alarm system vehicle orientations × 3 runs = 48	48	100
Tangent	B	2 vehicles × 2 speeds × 4 alarm system vehicle orientations × 3 runs = 48	48	100
Tangent	C	2 vehicles × 2 speeds × 4 alarm system vehicle orientations × 3 runs = 48	48	100
Tangent	D	1 vehicle × 1 speed × 4 alarm system vehicle orientations × 3 runs = 12	12	100
Tangent	E	2 vehicles × 2 speeds × 4 alarm system vehicle orientations × 3 runs = 48	48	100
Curved	A	1 vehicle × 1 speed × 2 curves × 2 alarm system vehicle orientations × 3 runs = 12	12	100
Curved	B	1 vehicle × 1 speed × 2 curves × 2 alarm system vehicle orientations × 3 runs = 12	12	100
Curved	C	1 vehicle × 1 speed × 2 curves × 2 alarm system vehicle orientations × 3 runs = 12	12	100
Curved	E	1 vehicle × 1 speed × 2 curves × 2 alarm system vehicle orientations × 3 runs = 12	12	100
Flagging	F	1 vehicle × 1 speed × 1 paddle condition × 3 runs = 3	3	100
Flagging	G	1 vehicle × 1 speed × 1 paddle condition × 3 runs = 3	3	100
Flagging	H	1 vehicle × 1 speed × 1 paddle condition × 6 runs = 6	6	100
Flagging	I	1 vehicle × 1 speed × 1 paddle condition × 3 runs = 3	3	100
Flagging	J	1 vehicle × 1 speed × 1 paddle condition × 3 runs = 3	3	100
Flagging	K	1 vehicle × 1 speed × 1 paddle condition × 3 runs = 3	3	100
Totals		273	273	100

Table 2. Worker Body Alarm Performance Testing Results

Testing scenario	Trajectory	Number of potential worker body alarm responses	Number of correct worker body alarm responses	Percent correct
Tangent	C	3 device locations × 48 runs - 15 data-collection equipment malfunctions = 129	129	100
Tangent	E	0 ft location × 48 runs = 48	48	100
Tangent	E	300 ft location × 48 runs - 3 data-collection equipment malfunctions = 45	45	100
Tangent	E	400 ft location × 48 runs - 6 data-collection equipment malfunctions = 42	31	74
Curved	C	3 device locations × 12 runs = 36	36	100
Curved	E	0 ft location × 12 runs = 12	12	100
Flagging	F	1 location × 3 runs = 3	3	100
Flagging	G	1 location × 3 runs = 3	3	100
Flagging	H	1 location × 6 runs = 6	6	100
Totals		324	313	97

Table 2 summarizes the results of the evaluation of the worker body alarm activations during the tests. Evaluated across the range of device locations, test scenarios, and vehicle trajectories, the worker body alarms achieved a 97% success rate. However, it was determined that all of the unsuccessful events for the worker body alarm were attributable to limitations in the effective communication range of the intrusion detection system with the worker body alarm. Ongoing improvements in the device are expected to significantly increase this communication range. All other device locations, scenarios, and trajectories tested achieved a 100% success rate.

Conclusions

A system test plan was developed to evaluate the accuracy and precision of the alarm system. The system successfully detected all actual intrusion threats as designed and performed in this test, and did not generate any false alarms under test conditions that were close to, but not actual, intrusion threats.

With regards to the operation of the worker body alarms, researchers found that the system also activated properly as long as the devices were in the effective range of the intrusion detection unit. Efforts continue to improve the effective range so as to further increase worker warning distances. It is recommended that field personnel also be reminded to monitor themselves and avoid straying farther away from the alarm system than its effective communications range with the intrusion detection unit.

It should be noted that the research team was unable to create a significant vertical curvature condition at its test facility. Consequently, it is not known how this factor would affect alarm system performance. Typically, it is recommended that work zones not be established just beyond the crest of vertical curves or beyond significant horizontal curves, and so the need for the system to protect under this condition is likely to be limited. The alarm system does offer significant opportunity to reduce work space intrusion risks to workers and motorists. However, field personnel should be regularly reminded that the system is designed to provide supplemental safety to work crews, and does not alleviate them of their responsibility to remain vigilant and ensure their own safety at all times. It is possible that field conditions not tested here may exceed the parameters for which the system was designed to handle. Field personnel should also be reminded to regularly check and maintain the system once it is fully deployed and in regular use.

Acknowledgments

This study was sponsored by Oldcastle Materials, Inc. The oversight provided by Curt Davison, Ken Dalton, and Nathan Creech, as well as Chad Green of ARTIS LLC, is gratefully acknowledged.

Author Contributions

The authors confirm contribution to the paper as follows: study conception and design: LT, GU; data collection: LT, TL; analysis and interpretation of results: LT, TL, GU; draft manuscript preparation: LT, GU. All authors reviewed the results and approved the final version of the manuscript.

References

1. Caltrans Division of Research, Innovation, and System Information. *Work Zone Intrusion Alarms for Highway Workers*. California Department of Transportation, Sacramento, CA, 2016. http://www.dot.ca.gov/newtech/researchreports/preliminary_investigations/docs/work_zone_warning_preliminary_investigation.pdf. Accessed April 5, 2016.
2. Ullman, G. L., M. D. Finley, and L. Theiss. *Work Zone Intrusion Countermeasure Identification, Assessment, and Implementation Guidelines*. Report No. FHWA/CA10-1102. Division of Research and Innovation, California Department of Transportation, Sacramento, CA, May 2010. http://www.dot.ca.gov/newtech/researchreports/reports/2010/final_report_task_1102.pdf.
3. Benekohal, R. F., and J. F. Linkenheld. *Evaluation of a Radar-Activated Horn System for Speed Control in Highway Maintenance Operations*. Report No. FHWA-IL-UI-235. Department of Civil Engineering, University of Illinois at Urbana-Champaign, IL, December 1990. https://apps.ict.illinois.edu/projects/getfile.asp?id = 2889.
4. Graham, J. L., J. Migletz, J. R. Loumiet, J. Hinch, D. Stout, and N. Lerner. *Maintenance Work Zone Safety*. Report No. SHRP-M/FR-89-001. Unpublished report from the Strategic Highway Research Program, National Research Council, Washington, D.C., 1989. http://onlinepubs.trb.org/onlinepubs/shrp/SHRP-89-001.pdf.
5. Krupa, C. *Work Zone Intrusion Alarm Effectiveness*. Report No. NJ-2010-004. New Jersey Department of Transportation, Trenton, NJ, September 2010. http://www.nj.gov/transportation/refdata/research/reports/NJ-2010-004.pdf.
6. Phanomchoeng, G., R. Rajamani, and J. Hourdos. *Directional Sound for Long Distance Auditory Warnings from a Highway Construction Work Zone*. Report No. CTS 08-20. Department of Civil and Mechanical Engineering, University of Minnesota, Minneapolis, MN, December 2008. http://www.its.umn.edu/Publications/ResearchReports/reportdetail.html?id = 1724.
7. Brown, H., C. Sun, and T. Cope. Evaluation of Mobile Work Zone Alarm Systems. *Transportation Research Record: Journal of the Transportation Research Board*, 2015. 2485: 42–50.
8. *A Policy on Geometric Design of Highways and Streets*, 4th ed. American Association of State Highway and Transportation Officials (AASHTO), Washington, D.C., 2001.

The Standing Committee on Work Zone Traffic Control (AHB55) peer-reviewed this paper (18-04225).

Article

TRR
JOURNAL OF THE TRANSPORTATION RESEARCH BOARD

Transportation Research Record
2018, Vol. 2672(16) 68–78

DOI: 10.1177/0361198118797461
journals.sagepub.com/home/trr
SAGE

Temporary Traffic Control Strategy Optimization for Urban Freeways

Laura E. Ghosh[1], Ahmed Abdelmohsen[2], Khaled A. El-Rayes[1], and Yanfeng Ouyang[1]

Abstract

Temporary traffic control (TTC) strategies have been widely adopted to maintain traffic flow while ensuring TTC costs remain a reasonable portion of construction budgets. As travel demand approaches the capacity of an existing facility, implementation of an appropriate TTC strategy is increasingly important not only because lane closures on these facilities exacerbate existing delay, but also because speeds associated with congestion contribute disproportionately and non-linearly to roadway emissions produced throughout the lifetime of the roadway. To enable stakeholders to make informed and transparent decisions on selecting a TTC strategy so as to balance the trade-offs among construction cost, traffic delay, and energy consumption, this paper discusses the development of an integrated model that identifies the Pareto-optimal front when construction start time, construction duration, lateral clearance, and width of shoulder borrowed as a through lane are considered as decision variables. A test implementation of the model suggests that when construction budgets are low, strategies for decreasing traffic delay differ significantly from those for decreasing energy consumption; however, as construction budgets increase, the objectives on traffic delay and energy consumption align much better.

Roadway construction accounts for 10% of annual congestion in the U.S., approximately 480 million vehicle hours of delay (*1*) and increased driver frustration as a result of unexpected slow-downs (*2*). As roadway network infrastructure ages and travel demand increases, the frequency of required maintenance and major rehabilitation projects also increases (*3*). Such activities typically require closing one or more traffic lanes to conduct work as well as to provide a buffer between the work site and open lanes. Temporary traffic control (TTC) strategies, also known as maintenance of traffic (MOT) plans delineate, with road markings and/or traffic cones, live lanes from longitudinal buffer, lateral buffer, and transition areas (Figure 1). They affect congestion through speed reductions, full or partial lane closure(s), live lane width reduction, and work intensity (workers and equipment). TTC strategies have therefore been developed to minimize construction-related travel delay while ensuring TTC costs remain a reasonable portion of construction budgets (*4*).

TTC strategies typically require additional costs and possibly increase construction project duration (*3*, *5*). Temporal TTC strategies take advantage of lower off-peak volumes by including overnight construction and implementing intermittent lane closures (to ensure full use of lanes during peak hours). Overnight construction reduces construction project duration, thereby reducing the number of peak hours affected by construction; however, additional lighting and labor costs are incurred. Intermittent lane closures, on the other hand, allow the roadway to operate at normal capacity during peak hours but increase construction project duration and require additional time to set-up and tear-down control devices (approximately 1 h each). Spatial TTC strategies mitigate construction-related congestion by increasing roadway capacity through use of freeway shoulders as through lanes, reduction of lateral clearance, and reduction of lane widths to accommodate an additional through lane. The use of shoulders requires additional construction time and costs for preparation. As lateral clearance and lane width reductions correlate to slower free-flow speeds through the construction site (*6*), these measures are expected to have diminishing returns in terms of construction-related congestion mitigation. Furthermore, decreasing lateral clearance between the

[1]Department of Civil and Environmental Engineering, University of Illinois at Urbana-Champaign, Urbana, IL
[2]Langan Engineering and Environmental Services Inc., Parsippany, NJ

Corresponding Author:
Address correspondence to Yanfeng Ouyang: yfouyang@illinois.edu

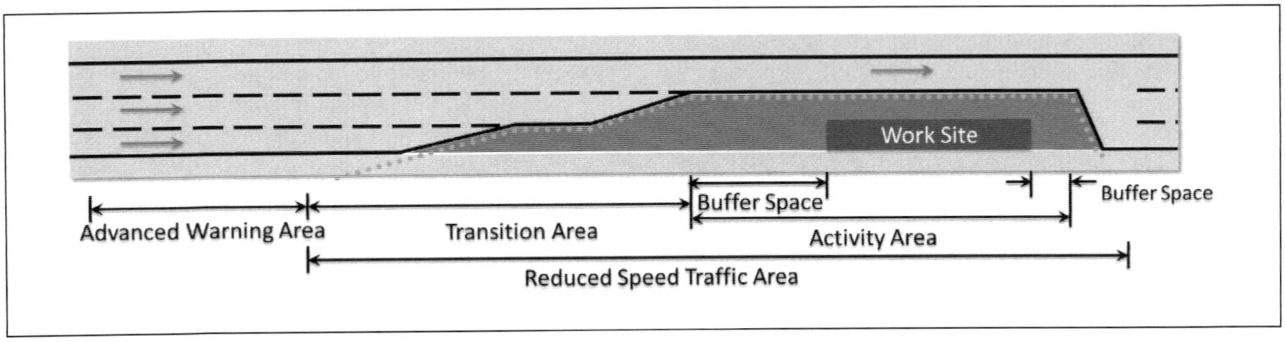

Figure 1. Typical elements of a TTC zone.

work area and live lanes requires more robust traffic barriers that are costlier and more time consuming.

Implementation of an appropriate TTC strategy as travel demand approaches the capacity of an existing facility is increasingly important not only because lane closures on these facilities exacerbate existing delay, but also because speeds associated with congestion contribute disproportionately and non-linearly to roadway emissions produced throughout the lifetime of the roadway. Several efforts have already been undertaken to model the effects of MOT plans in order to understand the trade-offs between traffic delay and construction costs. These efforts are not trivial, because of complex driver reactions to the adopted TTC measures, traffic composition, and their interactions. Possible modeling strategies range from macroscopic simulations that use average hourly traffic speeds and queue lengths (*7, 3*) to microscopic simulations that track the speed of each vehicle through the queue and work zone (*3*). In a program that identifies the optimal TTC strategy, both extremes have shortcomings: traffic delay estimates based on average hourly traffic speed and queue length do not fully account for the progression of traffic conditions along the roadway, and the computational burden of using microscopic simulations prohibits consideration of all possible TTC strategies.

In addition to traffic delay and construction costs, recognizing that approximately 23% of national greenhouse gas emissions are attributed to roadway vehicles (*8*), environmental goals concerning emissions are being incorporated into maintenance and rehabilitation decision making (*9, 10*). This school of literature typically focuses on reducing emissions from strategies such as the extraction/production, transport, and placement of construction materials (*11*). Emission impacts of the TTC activities are often simply estimated using vehicle miles of travel (VMT) (*12*).

Estimating emissions-related impacts of different TTC strategies is not straightforward. Unless traffic is diverted into alternate routes, VMT is not affected by the implementation of different TTC strategies. As such, emissions estimates of a TTC strategy should be analyzed based on vehicle hours traveled (VHT) at particular speeds. As traffic within a queue may not travel at a constant speed, and urban freeways may have multiple propagating queues initiating from multiple closely-spaced interchanges, it is imperative that an accurate TTC strategy analysis use speed estimates that do not obfuscate these speed reductions.

This paper introduces a methodology to identify the best TTC strategies that balance construction costs, traffic delay, and energy consumption trade-offs. Cost-benefit analyses that select a single optimal strategy may not accurately portray the trade-offs costs for two reasons: (i) emissions are not well enumerated for the purposes of selecting an optimal TTC strategy (*7*), and (ii) the true cost of emissions tends to be severely discounted by both users and agencies because of indirect and delayed effects on the quality of life. Instead, this paper focuses on the construction of an optimal Pareto front for TTC strategies which will enable stakeholders to make informed decisions that optimize these trade-offs. The following section presents the methodology used in developing the multi-objective optimization model. Next, a test implementation is used to illustrate the methodology and the expected results as well as their applicability are discussed. Finally, this paper concludes with some suggestions for future work.

Methodology

A multi-objective optimization model is constructed to evaluate and compare TTC design strategies in terms of traffic delay, energy consumption, and construction costs. The objective values produced by each of these modules are developed based on four decision variables that significantly impact the TTC strategies: (i) construction start time, T_{start}, (ii) daily duration of construction, T_{WS}, (iii) lateral clearance, W_{LC}, and (iv) width of shoulder borrowed for the live lane(s), W_{SH}. These decision variables are reported in previous studies to significantly affect construction costs and traffic delay (*2, 5, 6, 13–15*) and affect energy consumption (*16–19*). In particular, construction start time and daily duration of construction are selected to reflect decisions concerning temporal TTC strategies such as overnight and off-peak daytime construction. Meanwhile, as the number of closed lanes is initially assumed to be given, lateral clearance and width of shoulder borrowed as a through lane are selected to reflect decisions regarding spatial TTC strategies. As the minimum lane width is assumed to be 10 feet, lateral clearance, borrowed shoulder width, and the number of closed lanes are used to determine the width of open lanes.

Module 1: Traffic Modeling

Several previous work zone optimization studies have considered the effects of traffic on a macroscopic level by using average hourly traffic data, calculating a work zone capacity that corresponds to the number of lanes closed, and determining the queue delay resulting from decreased capacity in the work zone as well as moving delay through the work zone (*2, 6, 13, 20*). The methodologies employed in these studies hold true for sections of freeways in which queues do not extend to upstream interchanges, and are therefore applicable to rural and urban freeways when volumes are low and queues are short. However, urban freeways that are operating at or near capacity prior to construction may have queues extending beyond several interchanges. As the queue grows, traffic from upstream interchanges contributes to the queue, and within the queue area off-ramps allow for an increase in traffic speed and on-ramps may cause a secondary queue to form.

To properly reflect these traffic phenomena, Daganzo's cell transmission model (CTM) (*16, 21*) is employed. The CTM is a macroscopic traffic flow model that predicts the progression of traffic conditions along a roadway by discretizing the roadway into segments (cells) whose length is equal to the free-flow travel distance corresponding to a small amount of time (Δt). HCM 2010 freeway free-flow speeds may be used for non-construction segments, and work zone free-flow speed is affected by two of the decision variables: the number of closed lanes and lateral clearance, as shown in Equation 1 (*22*):

$$v_{\mathrm{ff\_WZ}} = 9.95 + 33.49\frac{v_{\mathrm{Posted}}}{v_{\mathrm{WZ}}} + 0.53v_{\mathrm{WZ}} - 5.6\frac{N}{{N_{\mathrm{OL}}}^2} - 3.84I_B - 1.71I_{\mathrm{DN}} - 1.45N_{\mathrm{Ramps}}, \quad (1)$$

where v_{Posted} is the non-construction posted speed;

v_{WZ} is posted speed in the construction area;

N is the number of open lanes under normal operating conditions;

N_{OL} is the number of open lanes in the work zone;

I_B is an indicator variable reflecting the barrier type and taking a value of 1 if cones or PE drums are used and 0 if a concrete barriers are used;

I_{DN} is a daytime or nighttime indicator variable taking a value of 1 if the construction occurs overnight and 0 otherwise; and

N_{Ramps} is the number of ramps within 3 miles upstream and downstream of the work zone.

Traffic progression into and out of highway cells is governed by a representative three-phase fundamental diagram that reflects either the roadway or construction zone free-flow speed, flow, and density. HCM 2010 freeway maximum capacity may be used for non-construction segments, whereas work zone capacity, $Q_{\mathrm{max\_wz}}$, is determined using Equation 2 (*23*):

$$Q_{\mathrm{max\_wz}} = 1857 - 168.1N_{\mathrm{CL}} - 37.0f_{\mathrm{CRL}} - 9.0P_{\mathrm{HV}} + 92.7W_{\mathrm{LC}} - 34.3T_{\mathrm{WS}}v_C - 106.1I_{\mathrm{WI}} - 2.3(G \times P_{\mathrm{HV}}), \quad (2)$$

where N_{CL} is the number of closed lanes;

f_{CRL} is an indicator variable taking a value of 1 if right-hand side lanes are closed and 0 otherwise;

P_{HV} is the percentage of heavy vehicles in the traffic stream;

W_{LC} is the width of the lateral clearance;

I_{WI} is a work zone intensity indicator variable taking a value of 1 if the activity is high intensity (i.e., bridge repair) and 0 otherwise; and

G is the grade of the roadway.

Per-lane jam density is assumed to be the same as the per-lane density under normal operating conditions. Therefore, the jam density within a cell is the per-lane jam density times the number of open lanes.

Using these cell properties, the CTM determines the number of vehicles transferred from an upstream cell, cell $i-1$, to the current cell, cell i, during the time interval Δt, $y_i(t)$, as given by Equation (3):

$$y_i(t) = \begin{cases} R_i(t) - r_i^{\mathrm{in}}(t), R_i(t) \leq \left(S_{i-1}(t) + r_i^{\mathrm{in}}(t)\right) \\ S_{i-1}(t), \textit{otherwise.} \end{cases}, \quad (3)$$

where $r_i^{\mathrm{in}}(t)$ is the number of vehicles entering the freeway at cell i during the time interval Δt;

$r_i^{\mathrm{out}}(t)$ is the number of vehicles exiting the freeway from cell i during the time interval Δt;

$S_{i-1}(t)$ is the number of vehicles able to be sent to cell i by the upstream cell, cell $i-1$, during the time interval $[t, t + \Delta t]$. It can exceed neither the number of available vehicles nor the maximum flow rate of that cell, Q_i; and

$R_i(t)$ is the number of vehicles able to be received by cell i during the time interval $[t, t + \Delta t]$. It governs the equation when the number of vehicles able to be received by cell i is less than the combined number of vehicles to be sent from the upstream cell, cell $i-1$, and the on-ramp and is given by Equation 4:

$$R_i(t) = min\begin{cases} \frac{w}{v_{ff}}[N_i - n_i(t)] + r_i^{out}(t) \\ Q_i \end{cases}, \quad (4)$$

where w is the speed of the backward shock wave;

v_{ff} is the free-flow speed of cell i;

N_i is the maximum number of vehicles able to occupy cell i at time t and is a function of jam density; and

$n_i(t)$ is the number of vehicles occupying cell i at time t, as given by Equation 5:

$$n_i(t) = n_i(t + \Delta t) + y_i(t + \Delta t) + r_i^{in}(t + \Delta t) - y_{i+1}(t + \Delta t) - r_i^{out}(t + \Delta t). \quad (5)$$

As each cell corresponds to a specified increment of time, Δt, total VHT is determined by summing the number of vehicles in each cell and multiplying by the time increment. Delay resulting from construction, the objective value produced from this module, is the difference in total VHT with and without construction.

Module 2: Energy Consumption

Fuel consumption is the objective value produced by Module 2 as a pseudo variable for emissions. Fuel consumption is suitable as it has a linear relationship with carbon dioxide emissions, which account for over 95% of transportation greenhouse gas emissions (*12*) and is readily calculable. Fuel consumption models based on typical driving patterns, such as the World Bank's Highway Development and Management Model (HDM-4) and the U.S. Environmental Protection Agency's Motor Vehicle Emission Simulator (MOVES) calculate fuel consumption for various vehicle types based on the power needed to overcome rolling, aerodynamic, inertial, and gradient resistances (*17*, *18*). As traffic composition is not expected to change unless traffic diverts onto alternate routes, many of the variables in both the HDM-4 and MOVES models can be treated as constants. The degree to which speed reductions result in increased/decreased emissions is unique for each vehicle type and is correlated to pavement roughness (*7*). Therefore, energy consumption regression equations for each vehicle type have been developed with VHT, average speed, and pavement roughness independent variables 6a, 6b and 6c (*16*, *19*):

$$\text{E} = VHT \times v \times \left(\frac{p}{v} + f_a + b \times v + f_c \times v^2\right), \quad (6a)$$

$$f_a = k_a \times IRI + d_a, \text{ and} \quad (6b)$$

$$f_c = k_c \times IRI + d_c, \quad (6c)$$

where E is estimated energy consumption (kJ);

v is an average speed (mph);

IRI is a measure of roadway roughness and stands for International Roughness Index; and

p, b, k_a, d_a, k_c, d_c, are model coefficients that vary for each vehicle type.

Equation 6a shows that energy consumption has a nonlinear relationship with speed. Empirical investigations reveal that slight reductions in traffic speeds (from over 70 mph to typical work zone speed limits of 45–55 mph) correspond to a reduction in emissions, whereas severely reduced speeds (typically below 35 mph) correspond to a significant increase (*10*, *24*). Such a relationship therefore requires careful categorization of the amount of VHT traveled at very low speeds, particularly if traffic encounters stop-and-go conditions. As such, VHT estimates from Module 1 are grouped by 5 mph increments and the average speed of each speed bin is used to estimate energy consumption.

Module 3: Construction Costs

Previous studies have sought to find the optimal work zone length through defining construction costs as the sum of set-up, relocation, and removal of traffic control devices, C_{TC}, highway construction costs, C_{HC} (*2*, *14*, *15*) and cost of preparing part of the shoulder to act as a through lane, C_{SH} (*13*). Only recently have the additional costs of nighttime lighting and overtime costs been incorporated into a comprehensive optimization program (*5*). This module accounts for set-up, relocation, and removal of traffic control devices, C_{TC}, highway construction costs, C_{HC}, worker overtime costs, C_{OT}, and nighttime construction costs, C_{NT}, as given by Equation 7:

$$\textit{Work Zone Construction Cost} = C_{TC} + C_{HC} + C_{SH} + C_{OT} + C_{NT}. \quad (7)$$

Here, for completeness of the paper and to highlight their relationship to the four decision variables, equations defining these costs are provided.

First, the costs associated with traffic control devices is a sum of the cost for their installation, C_I, use/rent, C_{TD}, relocation, C_{REL}, and removal, C_{REM}. All four traffic control devices costs are a function of the length of the work zone. Installation and removal costs are based on this single dependent variable as shown in Equation 8:

$$(C_I + C_{REM}) = (C_{Ii} + C_{REMi}) + T_{WS}v_C(C_{Iw} + C_{REMw}), \quad (8)$$

where T_{WS} is the daily duration of construction;

v_C is the speed of construction;

C_{Ii} is the fixed cost for the initial transportation and installation of traffic control devices at the beginning of the project such as those required for the warning, advanced, and termination areas;

C_{REMi} is the fixed cost for the removal and transportation of traffic control devices at the end of the project;

C_{Iw} is the per-mile cost of transporting and installing traffic control devices that are dependent on the length of the work site such as cones or barriers delineating travel lanes from the construction site; and

C_{REMw} is the per-mile cost of removing and returning the traffic control devices at the end of the project.

The cost to use/rent and relocate traffic control devices is also dependent upon the number of days or work sites needed to complete the maintenance/rehabilitation project. In this paper, it is assumed that there is one work site per day; therefore, the cost to use/rent and relocate traffic control devices is given by Equation 9:

$$(C_{TD} + C_{REL}) = N_{WS}[(C_{TDi} + C_{RELi}) + T_{WS}v_C(C_{TDw} + C_{RELw})], \quad (9)$$

where N_{WS} is the number of work sites needed to complete the maintenance/rehabilitation project;

C_{TDi} is the daily cost of using/renting the traffic control devices;

C_{RELi} is the cost of relocating the traffic control devices for the daily set-up and dismantling of the temporary work zone;

C_{TDw} is the per-mile cost of using/renting traffic control devices (as more cones/barriers will be needed for longer work sites); and

C_{RELw} is the per-mile cost of relocating the traffic control devices.

Note that there is a natural trade-off between the number of work sites required and work site length such that shorter work sites correspond to an increase in the number of work sites needed to complete construction.

Second, the cost to construct the highway, C_{HC}, is the baseline rehabilitation and maintenance costs, without implementing TTC measures designed to relieve congestion or reduce emissions, other than varying the length of the work site and number of work sites required to complete the project. These costs also include the cost of returning the roadway to a navigable condition, as this study assumes the construction area will be used as through lanes during non-construction periods. Thus, the cost to construct the highway is given by Equation 10:

$$C_{HC} = C_c T_{WS} v_C N_{WS}, \quad (10)$$

where C_C is the per-mile average construction cost.

Next, construction must begin by preparing part of the shoulder to act as a through lane. When the roadway has existing shoulders, preparation of the shoulder can range from \$2500 to \$100,000 per lane mile (*14*), with the lowest cost corresponding to locations where signing and striping are the only requirements and the highest cost corresponding to where the existing "shoulder must be removed, granular base adjusted and increased bituminous depth replaced" (*13*). The cost for preparing the shoulder, where necessary, is given by Equation 11:

$$C_{SH} = \begin{cases} C_{Shi} + (C_{ShR} \times W_{SH} \times (L_{Taper} + L)) & for\ W_{SH} > 0 \\ 0 & for\ W_{SH} = 0 \end{cases}, \quad (11)$$

where W_{SH} is the width of the shoulder and one of the decision variables;

C_{Shi} is the fixed mobilization cost to start work on the shoulder (\$);

C_{ShR} is the cost of preparing the shoulder for traffic (\$ per foot-mile);

L is the total length of the project (miles); and

L_{Taper} is the additional length needed to accommodate TTC tapering into the first work zone and out of the last work zone of the construction project.

Construction exceeding eight hours per weekday day and occurring on weekends incurs overtime costs (*25*). To compensate workers for exceeding their 40-hour work week or for working non-standard shifts, there is typically a premium cost for overtime hours worked. An equation for this premium cost as well as any additional overhead costs, C_{OT}, is given by Equation 12:

$$C_{OT} = f_{OT}\, C_C\, T_{WS}\, v_C\, N_{WS}, \quad (12)$$

where f_{OT} is the hourly cost adjustment factor for overtime premiums and additional overhead costs. This adjustment factor varies depending on work start time (to reflect non-standard shifts) and the number of overtime hours worked.

Finally, overnight construction has increased costs associated with required lighting equipment and nighttime safety devices; however, these costs are somewhat offset by reduced construction lag time resulting from improved material delivery times and faster cooling of asphalt material (*26*). The cost of overnight construction, C_{NT}, is therefore given by Equation 13:

$$C_{NT} = f_{NT}\, C_C\, T_{WS}\, v_C\, N_{WS}, \quad (13)$$

where f_{NT} is the hourly cost adjustment factor for nighttime construction costs.

Module 4: Multi-Objective Optimization

The primary purpose of this module is to find the Pareto-optimal set of strategies that provides a unique and non-dominated trade-off between construction costs, traffic delay, and emissions. Identification of the Pareto front provides decision makers the ability to change weights or utility functions without reconducting the multi-objective analysis (*27*). As the mathematical relationships between the objective values from Modules 1, 2, and 3 are not easy to determine, evolutionary algorithms such as genetic algorithms, ant colony optimization algorithms, and particle swarm optimization are best suited to conduct the multi-objective optimization (*28*). In keeping with other roadway construction-related multi-objective analyses (*5*, *29*, *30*) the non-dominated sorting genetic algorithm (NSGA-II) is adopted to estimate the Pareto front (*31*). NSGA-II is a genetic algorithm that uses a previous generation of strategies to construct a new diverse set of strategies that better approximates the Pareto front using a binary tournament selection operator that incorporates measures of optimality as well as diversity (*28*).

Case Study

To illustrate the proposed methodology and show the impact of considering energy consumption along with TTC implementation costs and traffic delay in a multi-attribute analysis, a case study was conducted to consider possible layouts for a single, inside lane closure for the purposes of roadway rehabilitation in 2011 along a 9.5 mile section (including Illinois 31 southbound exit through to Barrington Road entrance) of eastbound Jane Adams Memorial Tollway (I-90) in the greater Chicago Metropolitan Area (Project I-09-5588). This segment has three lanes plus, occasionally, one auxiliary lane to accommodate merging/diverging traffic as well as an inside shoulder with a width of approximately 12.5 feet and an outside shoulder with a width of approximately 11 feet. The posted speed limit is 60 mph.

Longitudinal dimensions for the TTC designs and corresponding speed limits were obtained from the Illinois Tollway standard lane closure drawings (*32*), with tapers of 800 feet followed by a 2,000 foot buffer before the next taper or work area. The tollway requires a minimum one foot of distance between the edge of pavement marking lines to the edge of the TTC device (*32*). Set-up and removal time for the TTC devices was assumed to require a total of 2 h. Finally, based on the construction schedule, 1 mile of maintenance is expected to require 20 h of work to complete. Thus the maintenance job would require at least 190 h of work.

Input Data and Assumptions

Module 1: Traffic Delay. The CTM considered a highway segment 4.18 miles upstream of the maintenance project (west of Randall Road) and 6.32 miles downstream (on-ramp to Illinois 53/I-290). This section of freeway has 14 ramps within 3 miles upstream and downstream of the construction segment. The corridor's calculated free-flow speed is approximately equal to 71 mph (*33*) and a time interval of 30 s was adopted. This corresponds to cell lengths of just under 0.6 miles under free-flow traffic conditions. A maximum flow of 2400 passenger vehicles per hour per lane and a jam density of 250 passenger vehicles per mile per lane were selected, resulting in a wave speed of 12 miles per hour. Under congested conditions, ramp traffic and mainline traffic were assumed to have equal priority when entering a downstream cell.

2012 mainline (*34*) and ramp (*35*) traffic volumes were adjusted to reflect a typical weekday in June (*36*). Commercial vehicles were calculated to comprise between 4 and 13.7% of the traffic along the corridor (*36*), of which 34.4% are light commercial vehicles, 40.2% are medium commercial vehicles, and 25.4% are heavy commercial vehicles. Traffic volumes in terms of passenger vehicles per vehicle per hour were calculated by applying a truck equivalency factor of 1.5 (*33*) to commercial truck volumes. It should be noted that traffic volumes on this segment are higher during the a.m. peak hour compared with the p.m. peak hour.

Without construction, the 20 mile study section incurs approximately 19,000 VHT per day, which corresponds to an average of 16.2 min of travel per vehicle.

Module 2: Energy Consumption. In 2024, roadway roughness on this segment is estimated to be 145.2 in/mile (2.29 m/km). Regression equations were developed for passenger vehicles and light, medium, and heavy commercial vehicles by our colleagues at the Illinois Center for Transportation of the University of Illinois at Urbana-Champaign, using January weekday climate conditions in Cook County, Illinois and an urban restricted access roadway type (*16*, *19*).

Without construction, energy consumption over this segment of freeway for a 24 h period is, on average, 4.94 gigajoules per day.

Module 3: Construction Costs. Construction costs were estimated using partial pay estimates for project I-08-5541 and adjusted to reflect the case study project. Construction activities are assumed to cost $1,150,000 per mile. The costs associated with traffic control devices used in the warning, advanced, and termination areas cost $2000 per day to rent, $25500 to install, $1000 to relocate, and another $1000 to remove. TTC devices such as cones and drums used to create the transition area and delineate traffic from the work and buffer area are assumed to cost $1000 per day to rent, $6000 per mile to install, $1000 per mile to relocate, and $2500 per mile to remove. Use of concrete barriers is assumed to cost an additional $1200 per mile to rent, $18,000 per mile to install, $40,000 per mile to relocate, and $20,000 per mile to remove. The cost associated with mobilization cost to start work on the shoulder is $950,000 and the cost of preparing the shoulder is $1300 per foot of shoulder width per mile.

The hourly cost adjustment factor for overtime premiums and additional overhead costs for non-standard work hours vary based on time of day such that compared with work conducted between 6:00 a.m. and 4:00 p.m., work conducted between 4:00 p.m. and 8:00 p.m. costs 50% more and work conducted between 8:00 p.m. and 6:00 a.m. costs double. Similarly, nighttime costs are assumed to increase by 20% between the hours of 6:00 p.m. and 6:00 a.m (*37*).

Module 4: Multi-Objective Optimization. This case study uses an initial population of 200 individual TTC strategies and ran for 15 generations. The fraction of each population that is made up of crossover children was limited to 20%. Mutation of a TTC strategy characteristic was allowed to deviate up to 0.1 standard deviations (scale ratio) and the rate at which the average amount of mutation was set to 0.5 (shrink ratio).

Results and Insights

An approximation of the Pareto front resulting from this analysis is shown in Figure 2. The figure shows that there is a fairly quick conversion to the Pareto front and that the Pareto front is well defined between $12.05M and $22M in construction costs, 0 to 305 thousand hours of traffic delay, and −3.14 GJ and 19.22 GJ of consumed energy. As can be seen in Figure 2*b*, there is a high correlation between energy consumption and traffic delay.

Figure 3 illustrates the Pareto front as a gray surface along with shaded values corresponding to the decision variables. Interestingly, all points on the Pareto front correspond to construction that starts at or before noon and ends at or before midnight. No Pareto front construction strategies correspond to work done only overnight. In addition, it should be noted that construction costs are very sensitive to the length of the segment cordoned off by TTC devices. In fact, the Pareto front has ruled out any segment length over 2.11 miles. To provide some perspective, Table 1 provides a comparison of key points in Figure 3.

Strategy A reflects the lowest construction cost ($12.05M) and highest traffic delay (179,000 hours of delay, which is an approximately 45% increase over non-construction conditions) and energy consumption (12 GJ, an approximately 11% increase over non-construction conditions). In comparison with strategy A, on the other hand, strategy F reflects the highest construction costs ($18.4M), a 50% increase compared with

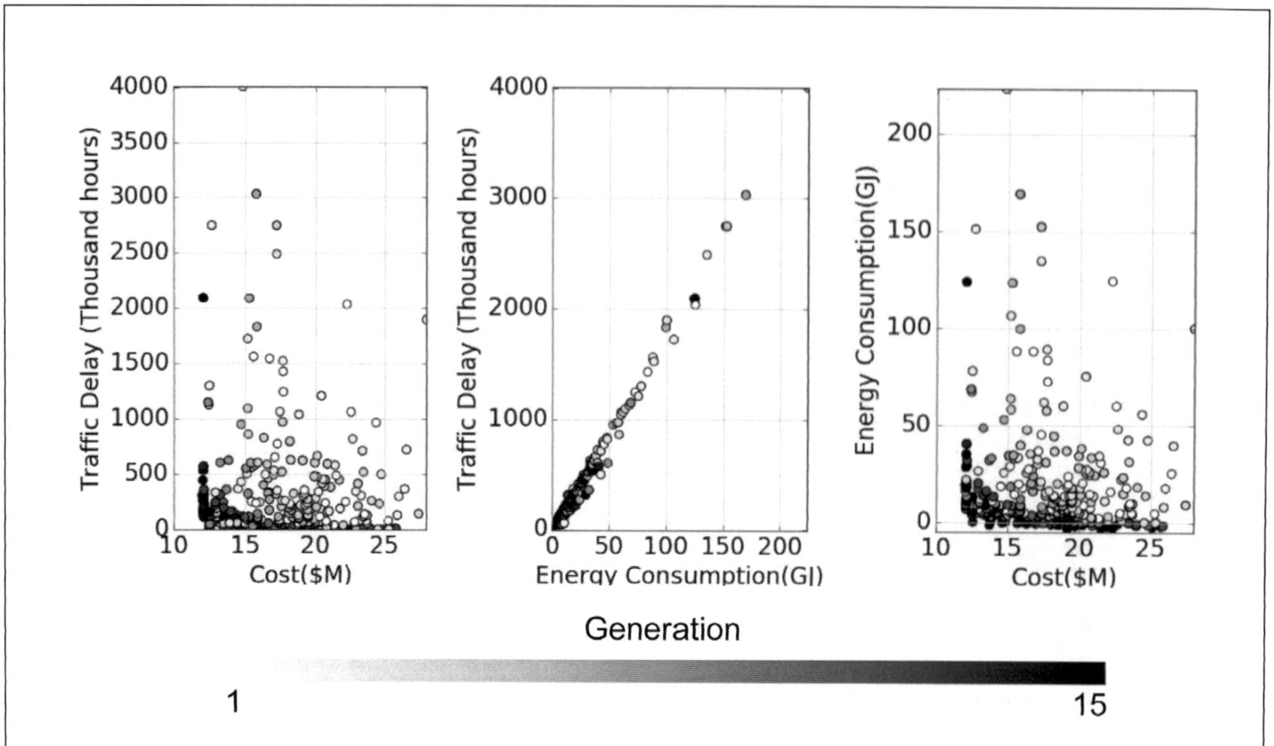

Figure 2. Pareto front approximation. (a) Cost vs. traffic delay; (b) energy consumption vs. traffic delay; and (c) cost vs. energy consumption.

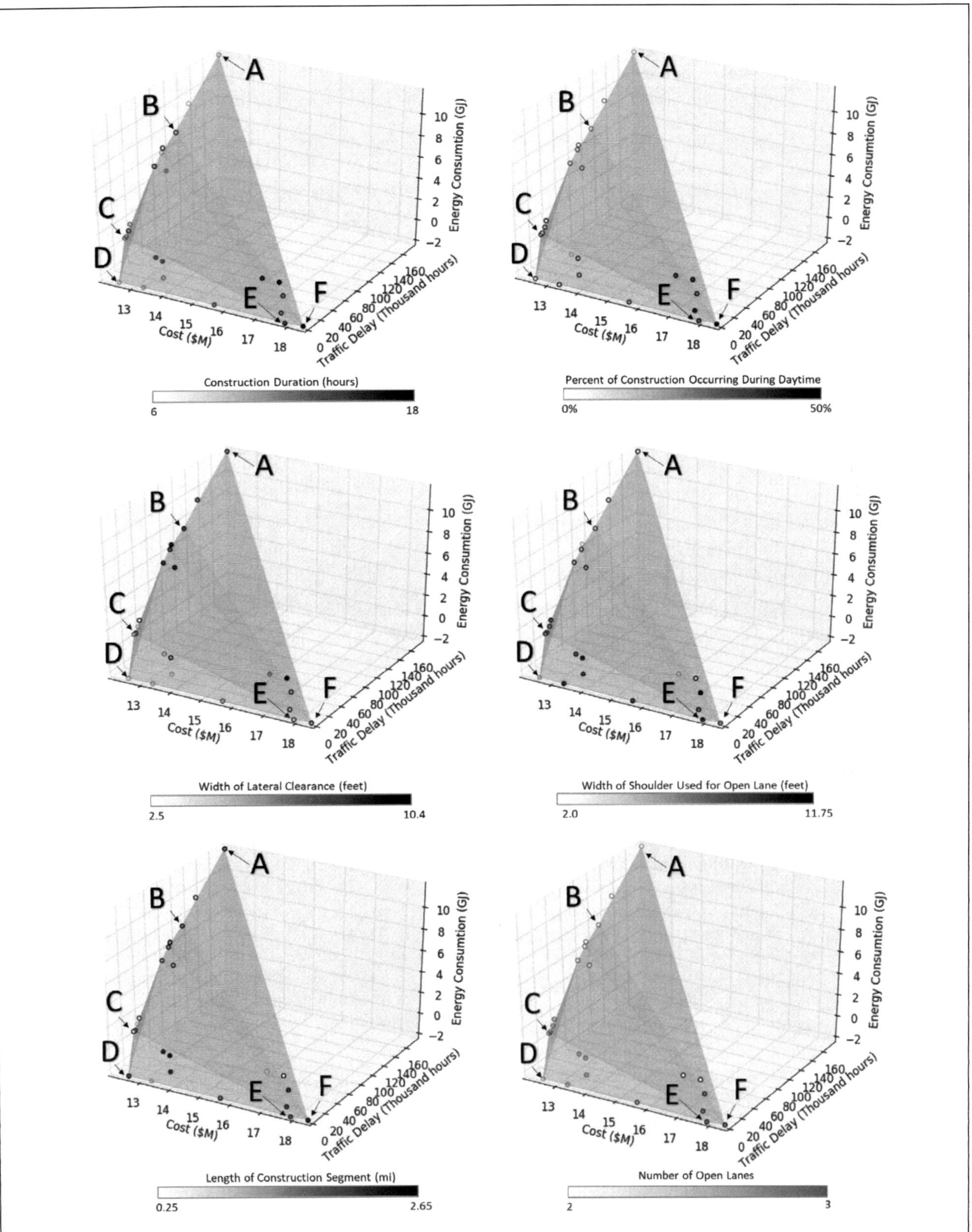

Figure 3. Investigation of the effects of decision variables on the Pareto front. Effects of: (a) construction duration; (b) percent of construction occurring during the daytime; (c) width of lateral clearance; (d) width of shoulder used for an open lane; (e) length of construction segment; and (f) number of open lanes.

Table 1. Summary of Optimal Work Zone Layout Decisions for Sample Strategies

Point	Start time	Work zone duration (hours)	Segment length (mi)	Lateral clearance (ft)	Open shoulder width (ft)	Number of open lanes	Cost ($M)	Delay (veh-hr)	Energy (GJ)
A	8 am	9	1.2	8.0	4.0	2	12.05	179k	12.0
B	10 am	6	1.1	9.5	2.0	2	12.06	109k	7.3
C	6 am	11	0.54	2.5	8.5	3	12.5	15k	1.4
D	6 am	11	1.46	4	10.5	3	12.6	0	−2.2
E	1 am	13	1.4	4	11	3	17.9	0	−2.2
F	Noon	12	2.1	4	10.5	3	18.4	0	−2.2

that of strategy A, no significant increase in traffic delay, and a decrease in energy consumption (−2.2 GJ, a 3% savings over non-construction conditions). Strategy D in this case study is the ideal solution, as the costs are only increased by 4%, traffic does not experience a significant increase in delay, and energy consumption is reduced by 3%. Figure 3 and Table 1 indicate that, for this case study, short duration construction zones that take advantage of daytime hours, utilize large lateral clearances, and use smaller amounts of shoulder width to provide three lanes of traffic (the name number of lanes as non-construction conditions) are low-cost strategies that effectively reduce traffic delay and energy consumption.

Conclusion and Future Work

To enable stakeholders to make informed decisions regarding traffic delay, energy consumption, and construction cost trade-offs when selecting a TTC strategy, this paper discusses the development of a model to identify the Pareto-optimal front using construction start time, construction duration, lateral clearance, and width of shoulder borrowed as a through lane as decision variables. After presenting the methodology used in developing the model, a test case was presented to illustrate how such a methodology can be implemented and the benefits of conducting such an analysis. Namely, the model generated a wide range of optimal trade-offs between traffic delay, energy consumption, and construction costs. Using these results, the model allows a decision maker with a fixed budget for a particular objective value to select an optimal TTC strategy. As in the test case, the model may also rule out several construction strategies (i.e., construction occurring solely overnight and long segments cordoned off by TTC measures were not ideal strategies), identify strategies that benefit one objective more than another, and identify transition points (i.e., approximately $12.5 million construction budget) beyond which relationships between the objective values and decision variables change.

Improvements to the model could include accounting for (i) the effects of energy consumption associated with shoulder widening and (ii) accounting for vehicular access to the construction site. Inclusion of such effects would further emphasize the difference between energy consumption and traffic delay. Furthermore, the model may be expanded to allow for additional objectives such as safety.

Author Contributions

The authors confirm contribution to the paper as follows: study conception and design: L. E. Ghosh, A. Abdelmohsen, K. A. El-Rayes, Y. Ouyang; data collection: L. E. Ghosh, A. Abdelmohsen; analysis and interpretation of results: L. E. Ghosh, A. Abdelmohsen, K. A. El-Rayes, Y. Ouyang; draft manuscript preparation: L. E. Ghosh, A. Abdelmohsen. All authors reviewed the results and approved the final version of the manuscript.

References

1. U.S. DOT. *Coordinating, Planning and Managing the Effects of Roadway Construction with Technology*. US DOT Intelligent Transportation Systems Joint Program Office, Washington, D.C., 2009. http://www.its.dot.gov/press/2009/WorkZone_QRv5%205.pdf. Accessed January 3, 2016.
2. Jiang, X., and H. Adeli. Freeway Work Zone Traffic Delay and Cost Optimization Model. *ASCE Journal of Transportation Engineering*, Vol. 129, No. 3, 2003, pp. 230–241.
3. Hardy, M., and K. Wunderlich. *Traffic Analysis Tools Volume IX: Work Zone Modeling and Simulation – A Guide for Analysts*. U.S. Department of Transportation Federal Highway Administration, Washington, D.C., 2009. http://ops.fhwa.dot.gov/wz/traffic_analysis/tatv9_wz/tatvol_9.pdf. Accessed January 16, 2016.
4. U.S. DOT Federal Highway Administration. *Manual on Uniform Traffic Control Devices for Streets and Highways 2009 Edition Including Revision 1 dated May 2012 and Revision 2 dated May 2012*. http://mutcd.fhwa.dot.gov/pdfs/2009r1r2/pdf_index.htm. Accessed January 3, 2016.
5. Abdelmohsen, A. Z., and K. El-Rayes. Optimal Trade-Offs Between Construction Cost and Traffic Delay for Highway Work Zones. *Journal of Construction Engineering Management*. *American Society of Civil Engineers*, Vol. 142, No. 7, 2016.

6. Benekohal, R. F., H. Ramezani, and K. A. Avrenli. *Queue and User's Costs in Highway Work Zones*. Final Report ICT-R27-33. Illinois Department of Transportation, Urbana, University of Illinois, IL, 2010.
7. Wang, T., I. S. Lee, J. Harvey, A. Kendall, E. B. Lee, and C. Kim. *UCPRC Life Cycle Assessment Methodology and Initial Case Studies for Energy Consumption and GHG Emissions for Pavement*. University of California Pavement Research Center, Davis, CA, 2012. http://www.its.ucdavis.edu/wp-content/themes/ucdavis/pubs/download_pdf.php?id=1862. Accessed July 31, 2015.
8. U.S. EPA. *U.S. Transportation Sector Greenhouse Gas Emissions: Fast Facts, 1990-2011*. Publication EPA-420-F-13-033. Office of Transportation and Air Quality, U.S. Environmental Protection Agency, Washington, D.C., 2013.
9. Egis. *Greenhouse Gas Emissions Mitigation in Road Construction and Rehabilitation: A Toolkit for Developing Countries*. The International Bank for Reconstruction and Development, The World Bank Group, Washington, D.C., 2011. http://www-wds.worldbank.org/external/default/WDSContentServer/WDSP/IB/2012/06/13/000425962_20120613142713/Rendered/PDF/696590ESW0P1010UBLIC00GHG0Web0final.pdf. Accessed January 11, 2016.
10. Bushman, R. J. *Probabilistic Performance Model for Evaluation of a Smart Work Zone Deployment*. Master's thesis. University of Saskatchewan College of Graduate Studies and Research Electronic Theses and Dissertations, Saskatchewan, Canada, 2007. https://ecommons.usask.ca/bitstream/handle/10388/etd-03192007-142727/Smart_Work_Zone_Performance_Model.pdf?sequence=1&isAllowed=y Accessed on July 19, 2016.
11. Smith, R., E. Ferrebee, Y. Ouyang, and J. Roesler. Optimal Staging Area Locations and Material Recycling Strategies for Sustainable Highway Reconstruction. *Computer-Aided Civil and Infrastructure Engineering*, Vol. 29, 2014, pp.559–571.
12. ICF Consulting. *Assessment of Greenhouse Gas Analysis Techniques for Transportation Projects*. American Association of State Highway and Transportation Officials Standing Committee on Environment, Washington, D.C., 2006.
13. Du, B., and S. Chien. Feasibility of Shoulder Use for Highway Work Zone Optimization. *Journal of Transportation Engineering*, Vol. 1, No. 4, 2014, pp. 235–246.
14. Schonfeld, P., and S. Chien. Optimal Work Zone Lengths for Two-Lane Highways. *Journal of Transportation Engineering*, Vol. 125, No. 1, 1999, pp. 21–29.
15. Chien, S., Y. Tang, and P. Schonfield. Optimizing Work Zones for Two-Lane Highway Maintenance Projects. *Journal of Transportation Engineering*, Vol. 128, No. 2, 2002, pp. 145–155.
16. Ghosh, L. E., L. Lu, H. Ozer, Y. Ouyang, and I. L. Al-Qadi. Effects of Pavement Surface Roughness and Congestion on Expected Freeway Traffic Energy Consumption. *Transportation Research Record: Journal of the Transportation Research Board*, 2015. 2503: 10–19.
17. Archondo-Callao, R. *Applying the HDM-4 model to strategic planning of road works*. The World Bank, Washington, D.C., September, 2008. https://www.illinoisvirtualtollway.com/Traffic/2012TrafficDataReport_Full_Small.pdfhttp://siteresources.worldbank.org/INTTRANSPORT/Resources/336291-1227561426235/5611053-1229359963828/tp-20.pdf. Accessed July 25, 2014.
18. *User Guide for MOVES2010b*. Document Catalogue Number: EPA-420-B-12-001b. Unite States Environmental Protection Agency, Washington, D.C., June, 2012.
19. Ziyadi, M., H. Ozer, S. Kang, and I. L. Al-Qadi. Pavement Roughness Related Energy Consumption and Environmental Impact Calculation Model for Transportation Sector. *Environmental Research Letters*, 2016.
20. Ramezani, H., and R. Benekohal. Analysis of Queue Formation and Dissipation in Work Zones. *Proc., 6th International Symposium on Highway Capacity and Quality of Service*, Stockholm, Sweden. 28 June–1 July, 2011. *Procedia Social and Behavioral Sciences*. Vol. 16, pp. 450–459.
21. Daganzo, C. F. The Cell Transmission Model, Part II: Network Traffic. *Transportation Research Part B: Methodological*, Vol. 29, No. 2, 1995, pp. 79–93.
22. Hajbabaie, A., C. Yeom, N. M. Rouphail, W. Rasdorf, and B. J. Schroeder. Freeway Work Zone Free-Flow Speed Prediction from Multi-State Sensor Data. Presented at 94th Annual Meeting of the Transportation Research Board, Washington, D.C., 2015.
23. Kim, T., D. J. Lovell, and J. Paracha. A New Methodology to Estimate Capacity for Freeway Work Zones. Presented at 80th Annual Meeting of the Transportation Research Board, Washington, D.C., 2001.
24. Grant, M., W. S. Hartley, R. Milam, J. Walters, L. O'Rourke, J. Brickett, and S. Sutter. *Handbook for Estimating Transportation Greenhouse Gases for Integration into the Planning Process*. Report No. FHWA-HEP-13-026. United States Department of Transportation, Federal Highway Administration, Washington, D.C., 2013. https://www.fhwa.dot.gov/environment/climate_change/mitigation/publications_and_tools/ghg_handbook/chapter05.cfm. Accessed January 7, 2016.
25. Larew, R. E. Are Any Construction Overtime 'Studies' Reliable? *Cost Engineering*, Vol. 40, No. 9, 1998, pp. 24–27.
26. Mostafavi, A., V. Valentin, D. M. Abraham, and J. Louis. Assessment of the Productivity of Nighttime Asphalt Paving Operations. *Journal of Construction Engineering and Management, American Society of Civil Engineers*. Vol. 138, No. 12, 2012, pp. 1421–1432.
27. Konak, A., D. W. Coit, and A. E. Smith. Multi-objective Optimization Using Genetic Algorithms: A Tutorial. *Reliability Engineering & System Safety*, Vol. 91, 2006, pp. 992–1007.
28. Ozcan-Deniz, G., Y. Zhu, and V. Ceron. Time, Cost and Environmental Impact Analysis on Construction Operation Optimization Using Genetic Algroithms. *Journal of Management in Engineering. American Society of Civil Engineers*, Vol. 28, No. 3, 2012, pp. 265–272.
29. El-Rayes, K., and K. Hyari. Optimal Lighting Arrangements for Nighttime Highway Construction Projects. *Journal of Construction Engineering and Management, American Society of Civil Engineers*, Vol. 131, No. 12, 2005, pp. 1292–1300.

30. Hyari, K., and K. El-Rayes. Optimal Planning and Scheduling for Repetitive Construction Projects. *Journal of Management in Engineering, American Society of Civil Engineers*, Vol. 22, No. 1, 2006, pp. 11–19.
31. Deb, K., S. Agrawal, A. Pratap, and T. Meyarivan. A Fast Elitist Non-dominated Sorting Genetic Algorithm for Multi-objective Optimization: NSGA-II. *Proc., 6th International Conference on Parallel Problem Solving from Nature*, 2000, pp. 849–858.
32. Illinois State Toll Highway Authority. *Standard Drawings Section E MOT*. March 31, 2016. http://www.illinoistollway.com/documents/10157/12283759/Section + E + - + MOT. Accessed July 31, 2016.
33. *Highway Capacity Manual 2010*. Transportation Research Board of the National Academies, Washington, D.C., 2010.
34. CDM Smith. *2012 Lane Closure Guide*. Illinois State Toll Highway Authority, Downers Grove, IL, 2013. http://www.illinoistollway.com/documents/10157/30214/ILLINOIS + TOLLWAY + LANE + CLOSURE + GUIDE_2010.PDF. Accessed July 25, 2014.
35. Illinois Department of Transportation. *Getting Around Illinois AADT Traffic Counts*. Illinois Department of Transportation, 2014. http://www.gettingaroundillinois.com/gai.htm?mt=aadt#. Accessed July 25, 2014.
36. CDM Smith. *2012 Traffic Data Report for the Illinois Tollway System*. Illinois State Toll Highway Authority, Downers Grove, IL, 2013. https://www.illinoisvirtualtollway.com/Traffic/2012TrafficDataReport_Full_Small.pdf. Accessed July 25, 2014.
37. Abdelmohsen, A. *Optimizing the Construction Planning of Highway Work Zones*. PhD thesis. University of Illinois Urbana-Champaign, Urbana, IL, 2016.

The Standing Committee on Work Zone Traffic Control (AHB55) peer-reviewed this paper (18-05061).

TRR
JOURNAL OF THE TRANSPORTATION RESEARCH BOARD

Article

Transportation Research Record
2018, Vol. 2672(16) 79–92

DOI: 10.1177/0361198118784143
journals.sagepub.com/home/trr
SAGE

Sun Glare: Network Characterization and Safety Effects

Danyang Sun[1], Karim El-Basyouny[1], and Tae J. Kwon[1]

Abstract

This research conducted a two-stage assessment to investigate sun glare effects on road safety in Edmonton. The first stage developed a methodology to model sun glare occurrence, aiming to identify when and where drivers are most likely to be exposed to sun glare. Safety risks at those identified locations during sun glare periods were assessed in the second stage. By contrasting the collisions during glare and non-glare conditions, this second stage aimed to quantify the effects of sun glare on road collisions. Consequently, three major findings were concluded. First, sun glare was found to significantly contribute to collision occurrence, especially at road intersections. Second, the effects of sun glare on collision occurrence during mornings on the eastbound and evenings on the westbound were observed to be particularly worse in the spring and fall months. In fact, safety in the southbound direction was significantly affected by sun glare during most daytime hours in the main winter months (November, December, and January). Lastly, the analysis revealed that certain collision types were more likely to occur during periods of sun glare. For example, collisions due to signal violations and failing to yield to pedestrians/cyclist were more likely to occur at intersections. At mid-block locations, the proportion of collisions occurring due to improper turning and lane changes were observed to be significantly higher. Overall, the approach proposed in this research provides a holistic method to quantify the effects of sun glare on road collisions and offers additional insights into the extent by which sun glare affects road safety.

Sun glare occurs when the sun is low in the horizon, usually an hour after dawn or before dusk. The direct sunlight exposure at a critical angle (normally less than 20° between the line of sight and the sunlight source) could result in a temporary dazzling sensation for drivers, which might impair their vision (*1*). This impairment could result in negative impacts on the safety of all road users. Although sun visors are installed in most vehicles, little protection can be provided when the sun is close to the immediate horizon and when it occurs unexpectedly after making a turn (*2*). According to the National Highway Traffic Safety Administration (NHTSA), sun glare is reported as the main cause of a few hundred collisions that occur each year in the U.S. (*3*). However, this figure is estimated to be under-reported as sun glare is rarely the sole cause of a collision and most drivers claim that they have been temporally blinded by sun glare. Another report from the Automobile Association in the U.K. shows that there were about 3,000 collisions related to sun glare blindness in 2013 (*4*). Compounding the problem is that glare times normally coincide with peak hours, making collisions even more likely to occur. Therefore, sun glare can be a significant safety concern and it requires additional research into its nature and its impact on road safety.

Although much previous research has shown that direct sunlight exposure is a significant causal factor that can impair visual performance, such as causing temporary blindness (*5–7*), there is only limited research examining and quantifying the sole effects of sun glare on road safety. Among those limited number of studies, Auffray et al. (*8*) found that the presence of sun glare affected the distribution of traffic flow, leading to a drop in speed and a loss of homogeneity among the vehicles. The authors further claimed that it was difficult to quantify the sole effects of sun glare and future research would, therefore, be necessary to investigate its direct impact on road collisions. Several other studies in Japan and the U.S. scrutinized the plausible relationship between collisions and sun glare presence, and found that sun glare significantly contributed to collision occurrence (*2*, *9*, *10*). However, several persistent issues were found, such as not controlling for confounding factors nor considering the influence of cloud cover. But perhaps the largest unaddressed issue is the inability to properly identify locations which are more likely to be exposed to sun glare due to the topography of the road network, and respective time periods when the effects of sun glare are going to be more prevalent due to the sun's position.

[1]Department of Civil and Environmental Engineering, University of Alberta, Edmonton, AB, Canada

Corresponding Author:
Address correspondence to Danyang Sun: danyang2@ualberta.ca

Acknowledging the need to address the above-mentioned limitations, this research aims to provide a better understanding of the safety effects of sun glare by combining a sun glare geographical assessment with a thorough collision analysis. Equally importantly, there were no previous studies into sun glare effects in Edmonton or across Canada. Nevertheless, the city of Edmonton, like most northern Canadian or European cities, is known to be one of the sunniest cities with more than 2,000 hours of bright sunlight each year according to the historical records from Environment Canada.

Thus, the objective of this paper is to evaluate the impact of sun glare on road safety in Edmonton, as a representative of most northern metropolitan cities. To achieve this objective, a two-stage analysis was conducted by first identifying the times (when) and locations (where) sun glare would likely occur, followed by a glare-related collision analysis where the focus was to examine the direct impact of sun glare on safety using the results obtained from the first stage.

Data

Several datasets (sun position, historical collision, and hourly weather data) were obtained and analyzed. Since large amounts of spatial datasets are to be assimilated, a Geographical Information System (GIS) platform, ArcGIS desktop 10.4 (*11*), was used to manage the data effectively.

GIS Data

A shape file containing Edmonton's road network was obtained from Office of Traffic Safety in Edmonton, and used as the base map. In addition, a digital elevation model (DEM) for Edmonton was also developed through the open data source provided by AltaLIS (*12*), which can be used to generate the shadow areas based on the terrain configuration. Analysis of the elevation profile revealed that most of Edmonton's terrain is quite flat and the areas with significant topographical changes are mainly located along the river valley, which hardly overlaps with the City's road network.

Sun Position Data

The position of the sun at a given time and location is usually defined in local coordinates by the azimuth angle and the elevation angle. The azimuth angle is the sun's relative positions along the horizon, measured clockwise from north ranging 0–360°. The elevation is the angle measured up from the horizon, with 0° at the horizon and 90° directly overhead of the terrain (*8*). These two local coordinates were extracted at an interval of five minutes throughout a typical year using an online tool available at the SunEarthTools website (www.sunearthtools.com) for Edmonton. The tool implements the Michalsky's algorithm (*13*) to precisely calculate the sun's positions for a specific location.

Collision Data

To investigate the sun glare effects on road safety, five years (2007–2011) of traffic collision data for Edmonton were used. The collision database is maintained by the City based on police reports. It includes detailed information for each collision such as time, location, collision types, and corresponding travel directions of at-fault vehicles. In this study, only collisions that occurred during weekdays were selected for the analysis to eliminate potential bias or differences in traffic flow behavior.

Weather Data

To identify sun glare occurrence, it is necessary to acquire detailed weather information to estimate the weather condition for each collision. In this study, the weather data for Edmonton were obtained from weather reports collected at Edmonton International Airport weather station from Environment Canada (*14*). Hourly weather data was extracted for each day from 2007 to 2011 to match the available time range of the collision data. The extracted weather data includes temperature, precipitation, visibility, and sky coverage condition.

Sun Glare Occurrence Analysis

To explore the effect of sun glare on road safety, it is important to understand how sun glare can be modeled. According to previous literature, the occurrence of sun glare can be determined by four main factors, which are sun's position, glare angle, terrain configuration, and weather (*8, 15*). To address the two most critical factors which are sun's position and glare angle (*10*), it is commonly accepted that the vertical and horizontal roadway geometries should point directly into the sun (*8*). The geometrical conditions can be illustrated as shown in Figure 1. As for the angle limit, a sun angle of 20° was acknowledged to cause glare in most studies (*1, 2, 8, 15*). This is also consistent with the findings from Zwaheln (*16*), which suggested the maximum angular distance for one's visibility to be normally 20°. Therefore, an angle limit of 20° was chosen for both the horizontal and vertical geometries in this research. A potential glare time window was first identified to model the vertical angle of glare based on the extracted sun elevation data. With the identified glare time windows, the corresponding sun glare risk locations were then identified by considering the horizontal geometry. Since Edmonton's terrain is quite flat, it is expected that the terrain configuration will not affect the final results in any meaningful way. Nonetheless, a Hillshade map was introduced in this study to account for shadow effects (*17*). Lastly, the influence of weather factors was considered as outlined in the collision analysis.

Sun Glare Time Window Selection

Two glare time windows, morning and evening glare, were identified based on the sun's elevation for each month over a typical calendar year. The glare time window of each month was calculated by finding the intersection of the time intervals with low sun elevation between 0 and 20° of each day in the month. This method ensures that the identified glare time window can represent a consistent condition for the sun for the whole month. The resulting time windows are shown in Table 1. Once the glare time windows were identified, the corresponding solar azimuth ranges (sun cone) for each month were extracted using a similar algorithm to find the intersecting parts of each day. It was found that during the three main winter months (November, December, and January) the sun always had an elevation under 20° during all daytimes. To further clarify the sun glare effects during winter, the winter glare time periods were split according to the cardinal direction of the sun. The solar azimuth angle was broken into in four 90° orientation quadrants. The time periods with sun azimuths in the East quadrant (azimuth from 45 to 135°) were identified as the eastbound glare period. Accordingly, southbound and westbound glare periods were also identified. The identified solar azimuth is also shown in Table 1.

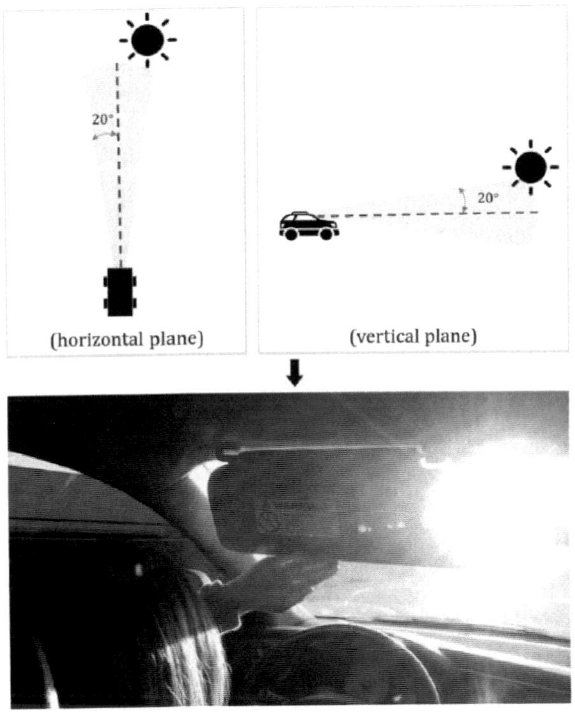

(Source: https://www.flickr.com/photos/mrmatt/2141229491)

Figure 1. Geometrical conditions of sun glare.

Sun Glare Prone Location Identification

With the previously defined sun glare time, the corresponding locations were identified by using ArcGIS. There are two main phases in the identification process. The first phase was carried out to address the angle of glare from a horizontal geometrical perspective by matching the road network alignment with the sun's position. To achieve this, three steps were carried out including changing road segments from polylines to single lines; calculating the road bearings; and carrying out a road searching within the 20° to the azimuth sun cone. The second phase was to address the factor of terrain configuration and exclude road segments in shadow areas. A Hillshade tool of ArcGIS was employed to model

Table 1. Typical Sun Glare Time Windows for Edmonton

Feb-Oct	Morning glare		Evening glare		Eastbound glare		Southbound glare		Westbound glare	
Month	Time window	Sun azimuths	Time window	Sun azimuth	Time window	Sun azimuth	Time window	Sun azimuth	Time window	Sun azimuth
February	8:20–10:00	112.3–139.5°	15:30–17:15	223.8–241.7°	–	–	–	–	–	–
March$_{1–7th}$	7:20–9:45	99.0–132.1°	15:45–18:15	228.9–259.1°	–	–	–	–	–	–
March$_{15–31st}$	7:45–9:30	88.6–113.1°	17:45–19:40	249.6–268.3°	–	–	–	–	–	–
April	7:05–8:30	74.7–97.7°	18:45–20:10	270.2–279.6°	–	–	–	–	–	–
May	6:00–7:45	58.8–83.5°	19:15–21:05	280.4–297.9°	–	–	–	–	–	–
June	5:10–7:45	48.5–78.8°	19:30–21:50	283.3–309.0°	–	–	–	–	–	–
July	5:50–8:00	54.28–83.0°	19:30–21:30	282.6–302.3°	–	–	–	–	–	–
August	6:45–8:15	68.42–93.4°	19:00–20:20	272.9–284.3°	–	–	–	–	–	–
September	7:30–9:00	84.4–111.5°	18:00–19:10	255.4–264.9°	–	–	–	–	–	–
October	8:30–10:00	105.19–131.6°	16:30–18:00	230.9–246.3°	–	–	–	–	–	–
January	–	–	–	–	8:50–9:30	129.5–135.6°	9:45–16:00	140.7–225.3°	16:00–16:25	225.3–230.5°
November $_{8–30th}$	–	–	–	–	8:30–9:15	127.4–137.1°	9:15–15:30	137.1–222.6°	15:45–16:20	225.7–232.4°
December	–	–	–	–	–	–	9:00–16:00	132.7–226.1°	–	–

Note: Effects of daylight saving time change on the position of the sun were eliminated by excluding the days of March 8–14th and November 1–7th.
– = not applicable.

the sunlight illumination of a surface to identify the shade areas on a DEM raster. Once the sun coordinates are entered, the Hillshade tool calculates the illumination value for each cell in the raster, which can be used to develop shadow maps for the use of erasing covered sun glare prone locations. Although only terrain configuration was examined in this study, the methodology described here proposes a way for future research to further address the shadow influence by surface objects if a more detailed DEM is available. Besides, the downtown core area of Edmonton where tall buildings are mainly located was excluded from this analysis since the shadow from the buildings could shade some of the roads.

As a result, a series of visualization maps were created for each glare time window and its corresponding glare prone locations as shown in Figure 2. The percentage of total length of sun glare prone road segments was also calculated and labeled on each map. As most parts of the road network in Edmonton are arranged in typical east–west/north–south grids, the distributions of identified locations on the maps during morning and evening glare periods were quite similar for most of the months and consisted of roads with east–west orientations. This is due to the fact that sun glare normally occurs right after sunrise in the east or before sunset in the west. The coverage of sun glare affected roads in these two glare periods accounted for about 40% of the total road network length, with two exceptions being the months of February and October, which only had a coverage percentage from 11.9% to 14.4%. As for the other three months in winter, the sun glare was found to occur throughout the entire daytime. However, it was found that during most of the glare times in winter, the sun was facing south, meaning that the majority of exposed roads were those with north–south orientations.

Assessment of Sun Glare Impacts on Collisions

To directly investigate sun glare effects on road safety, historical collisions were examined and contrasted between glare and non-glare conditions.

Glare Collision Data Extraction and Assembly

In this stage, historical collision data with the potential of sun glare involvement was extracted and assembled for examination. For each month, collisions that fell in the glare time windows and occurred at the identified locations were selected. A case-control design method was used to control for potential confounding factors. To exclude the effects of other adverse weathers, only two weather conditions, clear and overcast were examined. The case group and control group were developed based on the observed collisions under the two weather conditions, respectively. As vehicles with different travel directions on the same road have different sunlight exposure, the collisions in each group were split according to the travel direction of the vehicle at fault.

By examining the sun position data for Edmonton, it is apparent that eastbound traffic would be exposed to sun glare in the morning glare period and westbound traffic would be exposed in the evening glare period. Except for the case with sun glare exposure controlled by the cloud cover, the control group were found to be comparable to the case group in terms of time of day and location. Therefore, it can be expected that the difference of collision frequencies among different directions is caused by the exposure to the sun's glare. To further exclude the unrelated collisions, the Michalsky's (*13*) sun position algorithm was applied on each collision to calculate its specific sun position. As a result, any collisions where the road or approach bearing differs above 20° compared with the sun azimuth were removed from the analysis. The final extracted data is shown in Table 2. It should be noted that there were only two directions (east and west) for mid-block crashes, as only the roads with east–west orientation matched the sun azimuth.

Overall Effects of Sun Glare on Collision Occurrence

The aggregated collision data in Table 2 from February to October over the five analysis years was used to examine the general sun glare effects. Collisions occurring in the other three winter months (November, December, and January) were analyzed separately in the next section due to their unique sun positions. Three statistical techniques including the Chi-square test of independence, Configural Frequency Analysis, and the Wilcoxon signed-rank test were used and their results were summarized below.

Results of the Chi-Square Test of Independence. The Chi-square test of independence was used to determine the relationship between sun glare and its impact on collision frequency. The two categorical variables in the contingency table of each period were set as the cloud cover conditions (clear or overcast) and the collision direction (exposed to sun glare or not). If there is a significant association between these two variables and the increase of collisions falls in the directions exposed to sun glare, it can be interpreted that the sun glare has an adverse impact on the road safety. Table 3 shows the results of the chi-square test of independence.

As can be seen, for intersection collisions, a statistically significant association was found in both morning and evening glare periods using a 95% confidence level. To check if sun glare increases the collisions as hypothesized, an odds ratio (OR), which compares the odds of exposed collisions among cases to the odds among controls, was computed for the glare exposed directions. It was found that the odds ratios for both an eastbound direction in the morning and westbound direction in the evening were larger than one, meaning

Figure 2. (*a*) Glare prone locations of morning glare periods, (*b*) glare prone locations of evening glare periods, and (*c*) glare prone locations of January, November, and December.

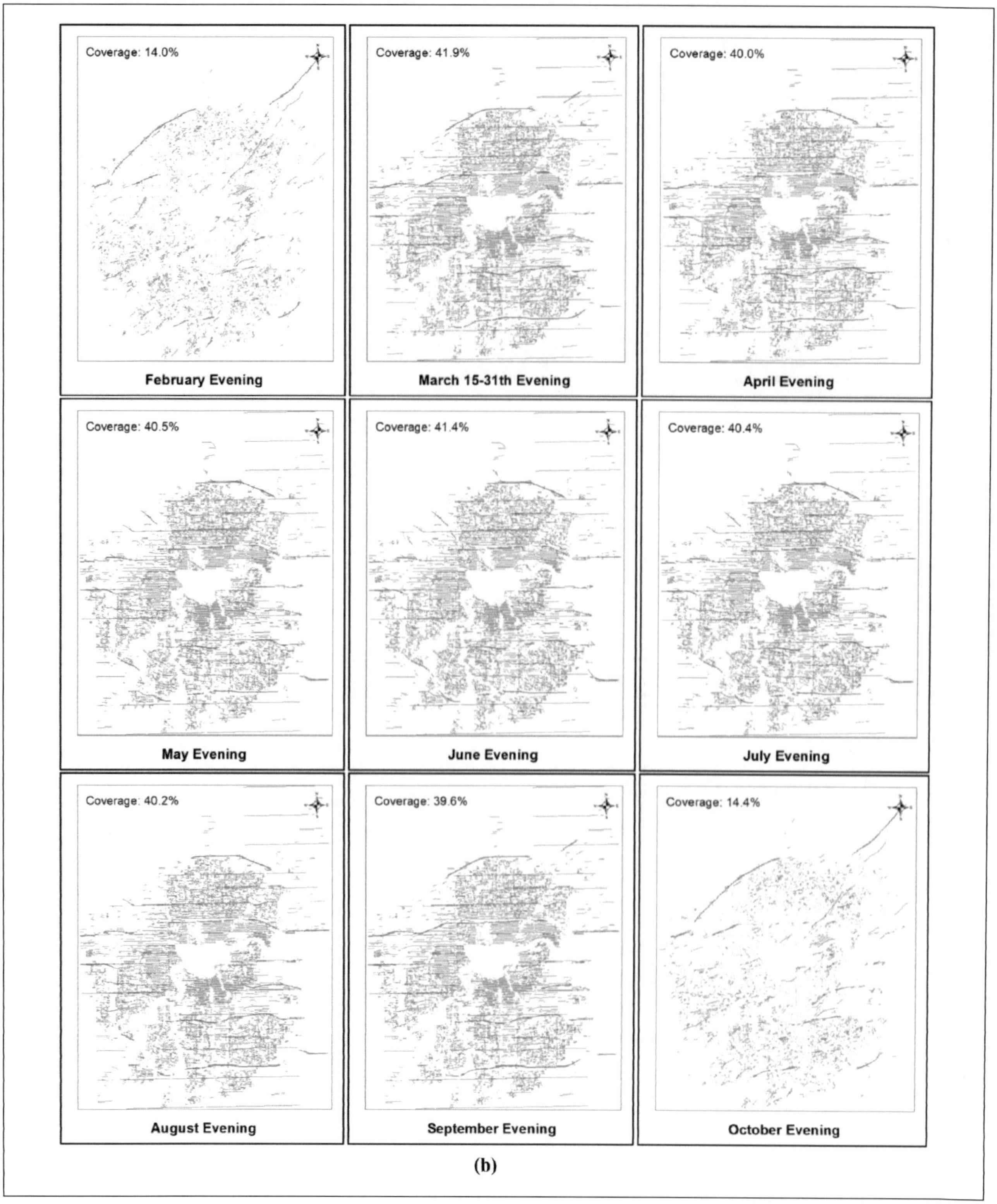

Figure 2. (Continued)

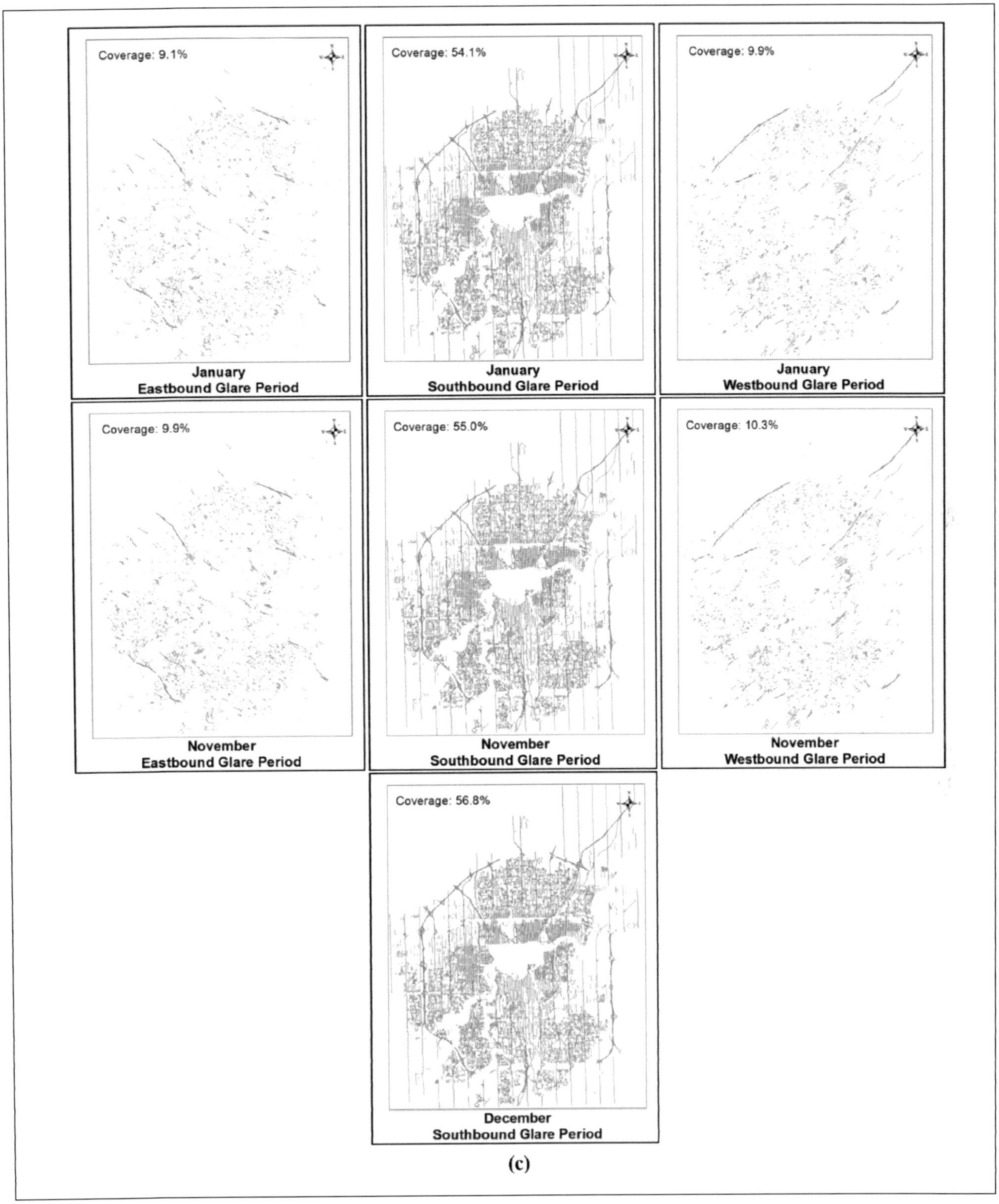

Figure 2. (Continued)

Table 2. Observed Collisions over Different Glare Periods and Directions

Location type	Glare periods	Months	Group	Travel directions East	West	North	South
Intersection	Morning glare	Feb–Oct	Case	258	224	184	208
			Control	250	245	290	264
	Evening glare	Feb–Oct	Case	262	241	251	205
			Control	294	227	328	257
	Winter all daytime glare	Nov, Dec, Jan	Case	360	343	348	400
			Control	499	483	530	486
Mid-block	Morning glare	Feb–Oct	Case	227	181	–	–
			Control	200	222	–	–
	Evening glare	Feb–Oct	Case	132	144	–	–
			Control	152	182	–	–
	Winter all daytime glare	Nov, Dec, Jan	Case	5	4	194	132
			Control	8	10	274	180

Note: – = not applicable.

Table 3. Results of Chi-square Test for Overall Effects

Location type	Glare periods	Group	Observed collisions Exposed direction	Observed collisions Non-exposed direction	Expected collisions Exposed direction	Expected collisions Non-exposed direction	χ^2	$P(\chi^2)$	$OR\left(\frac{a*d}{b*c}\right)$
Intersection	Morning	Case	258 (a)	617 (b)	231	644	7.98	<0.01	1.340
		Control	250 (c)	799 (d)	277	772			
	Evening	Case	241 (a)	718 (b)	217	742	6.27	<0.01	1.301
		Control	227 (c)	880 (d)	251	856			
Mid-block	Morning	Case	227 (a)	181 (b)	210	198	5.94	<0.01	1.404
		Control	200 (c)	222 (d)	217	205			
	Evening	Case	132 (a)	144 (b)	128	148	0.32	0.57	1.097
		Control	152 (c)	182 (d)	155	179			

that the exposure to sun glare increases the crash risk at intersections. In fact, collisions increased by 34.0% on the eastbound direction during the morning and 30.1% on westbound during the evening.

The results pertaining to the mid-block related collisions indicate that sun glare had a significantly higher collision risk during the morning glare, approximately 40% more collisions when compared with the opposing direction, at the 95% confidence level. An increase of 9.7% was found during the evening glare; however, this change was not statistically significant.

Results of Configural Frequency Analysis. To further examine the sun glare impact on collisions, a Configural Frequency Analysis (CFA) was also employed in this study. The goal of this method is to detect types and antitypes in the contingency table of the collision data assembled in this study. Types are defined as patterns of variable categories that occur significantly more than expected by chance, while antitypes are "significantly less," both of which are identified by comparing observed and expected frequencies in individual cells (configuration patterns). From these detected patterns, the researchers can interpret whether certain variable values in different categories generally occurred together and are statistically related from a person-oriented perspective. To perform a CFA, two main routine steps were conducted as follows: i) a first order CFA model was used as the base model to calculate the expected cell frequencies under an a priori assumption that no variable relationship exists (*18*); ii) compare the observed and expected frequencies using a statistical testing. In this study, *Z* test was employed to detect the types and antitypes. Other tests may also be used. For more details on CFA, readers may refer to the work by Eye et al (*18*).

In this section, the CFA was conducted to compare the collisions among different travel directions for glare and non-glare conditions. The null hypothesis was that the sun glare occurrence was independent from the collision direction, meaning the collision distributions on different directions should be consistent regardless of cloud cover (i.e., clear or overcast) in the two groups. Therefore, if it is possible to detect a type (i.e., significantly greater number of

Table 4. Results of Configural Frequency Analysis for Overall Effects

Location type	Glare periods	Group	Travel direction	Collision frequencies			Statistics		
				Observed(o)	Expected(e)	o − e	z	P(z)	Type/antitype
Intersection	Morning	Case	East	258	231	27	1.91	0.03	type
			West	224	213	11	0.80	0.21	
			North	184	216	−32	2.27	0.01	antitype
			South	208	215	−7	0.49	0.31	
		Control	East	250	277	−27	–	–	–
			West	245	256	−11	–	–	–
			North	290	259	31	–	–	–
			South	264	257	7	–	–	–
	Evening	Case	East	262	258	4	0.24	0.40	
			West	241	217	24	1.70	0.04	type
			North	251	269	−18	1.15	0.13	
			South	205	214	−9	0.71	0.24	
		Control	East	294	298	−4	–	–	–
			West	227	251	−24	–	–	–
			North	328	311	17	–	–	–
			South	257	248	9	–	–	–
Mid-block	Morning	Case	East	227	210	17	1.40	0.08	type
			West	181	198	−17	1.43	0.08	antitype
		Control	East	200	217	−17	–	–	–
			West	222	205	17	–	–	–
	Evening	Case	East	144	148	−4	0.33	0.37	
			West	132	128	4	0.34	0.37	
		Control	East	182	179	3	–	–	–
			West	152	155	−3	–	–	–

Note: – = not applicable.

observed collisions than expected) for the collisions on the glare-exposed directions in clear weather conditions, the null hypothesis is rejected and it can be concluded that the sun glare plays a significant role in increasing collision frequencies. If an antitype is detected for the directions not exposed to sun glare in the clear weather, the findings of adverse effects of sun glare will be even stronger. The type and antitype can also be detected in the overcast weather conditions; however, this would just reflect the reversing of the results in clear weather as there are only two weather scenarios in the current CFA test. The focus of this study is to investigate the types/antitypes in the clear weather conditions since the primary concern is to determine whether collision frequencies would increase when the visibility is impaired by the sun.

Table 4 summarizes the results of CFA for testing the overall sun glare effects. For intersection collisions, two types were detected in the case group during clear weather conditions at 95% confidence level, which corresponded to the case of sun glare occurrence. While for mid-block crashes, only one significant type (at 90% confidence level) was found in the eastbound direction during the morning glare period. No other significant types were found. The result obtained from CFA is similar to that of the Chi-square test, which indicates the safety of intersection traffic is significantly affected by sun glare in both the morning and evening, while the safety of mid-block traffic seems to be only affected during the morning. The possible reason for the differences of mid-block collisions between morning and evening is that drivers are more likely to be in a hurry in the morning on workdays to get to work. Such a finding cannot be substantiated from the analysis in this study since speed information was not available. However, analysis showed that the significance of mid-block morning collisions was not as high as intersection collisions. Therefore, it can be generally concluded that the adverse effects of sun glare on intersections are greater compared with mid-blocks.

Results of the Wilcoxon Signed-Rank Test. Since the data for the same month in the case and control groups can be considered as matched pairs controlled by the differences in sun glare exposure, the Wilcoxon Signed-Rank Test (W test) could be used. It is an alternative to the paired t-test, which can be applied when the samples do not follow a normal distribution such as the collision frequencies. In this study, the comparison was conducted using the ratio of collisions on glare exposed direction to non-glare exposed directions between the two groups. The ratio X_i was paired for each month of the groups and can be calculated as shown in Equation 1, where the glare exposed directions are termed as east in the morning and west in the evening.

Table 5. Results of Wilcoxon Signed-Rank Test for Overall Effects

Collision type	Number of pairs	Average glare collision ratio X_i		W-statistic	z	P(z)
		Case group	Control group			
Intersection collisions	18	0.38	0.30	123	2.91	<0.01
Mid-block collisions	18	1.10	0.99	35	0.76	0.22
Overall	36	0.74	0.64	183	1.49	<0.1

$$X_i = \frac{\textit{crashes on glare exposed direction}}{\textit{crashes on nonglare exposed direction}} \quad (1)$$

The null hypothesis of this test was that there was no difference between the two groups of collision ratios, while the alternative hypothesis was that this kind of ratio would be higher under the occurrence of sun glare. The corresponding steps and formulas of this test can be found in the work by Siegel (*19*).

The result summary is shown in Table 5. For intersection collisions, a significantly higher odds ratio X_i of collisions along glare direction was found in the case group compared with the control group, indicating that sun glare increases collision occurrence along the glare directions. As for mid-block collisions, no significant differences were found between the two groups. When intersection collisions and mid-block collisions were combined, it was found that the ratio X_i was still significantly higher in the case group, which supports the adverse effect of sun glare on safety in general. The result of the *W*-test confirms the findings of the two previous tests and shows further evidence that the sun glare increases the risk of collisions, especially at intersections.

Monthly Glare Collisions Analysis

To examine this monthly variation of sun glare effects, two types of analysis are explored in this section. One is a series of CFA tests for each month from February to October, during which the morning and evening glare periods are distinct. The other one is a separate analysis for the 3 winter months (November, December, and January) when the sun is always low on the horizon throughout the daytime.

Monthly Analysis for February through October. A series of CFA tests were performed for each month based on the comparison between clear and overcast weather in the same way as the overall analysis. Table 6 shows the results. For intersection collisions, one type was found in the month of September during the morning glare at a 90% confidence level. The other type was found in the month of March during the evening glare at a 90% confidence level. As for mid-block collisions, the results indicated that March had a significantly higher number of observed collisions along the eastbound direction in the morning period, while there was no significant difference detected between observed and expected collisions in the evening period. From these results, it can be concluded that March and September are the two months that are more likely to be affected by sun glare. Since the two months are around the Spring and Autumn Equinoxes, it can be inferred that the sun vision obstruction on Edmonton's grid road network is especially potent when the sun rises exactly in the east and sets exactly in the west. Another possible reason is that the morning and evening glare time periods in March and September overlap with the peak traffic hours, which indirectly increases the risk of collision due to increased traffic volume.

Winter Monthly Analysis. For winter monthly analysis, the collision numbers were counted during each of the winter glare time windows as previously defined in Table 1. Afterwards, the winter glare effect was examined using CFA for the three defined glare periods (eastbound glare period and southbound glare period and northbound glare period). The odds ratios OR of collisions along different directions were also computed by comparing the glare and non-glare conditions. The result summary is shown in Table 7.

At intersections, the southbound collisions during the southbound glare period were found to be significantly higher than expected (at 95% confidence level) in the three months. The corresponding OR indicated that the southbound collisions were increased by 23% when sun glare occurs from that direction. More than expected numbers of intersection collisions were also recorded in clear weather along the eastbound and westbound directions during the eastbound and westbound glare periods, respectively. However, these two differences were not statistically significant. This could be attributed to the small sample size in these two periods. While on mid-blocks, since the number of observations was too few in the eastbound and westbound glare periods, the CFA was not conducted. Only the result for the southbound glare period is presented in the table. Similar to previous tests, no significant type was detected for any directions.

Effects of Sun Glare on Specific Collision Types

Although the above results showed that sun glare had a generally adverse effect on road safety, it is important to examine what types of collision are more likely to be affected due to sun glare vision impairment. Therefore, a CFA test was

Table 6. Results of Monthly Analysis from February through October

Glare periods	Month	Glare collisions			Statistics		Type/antitype
		Observed (o)	Expected (e)	o − e	z	P(z)	
Intersection morning glare	Feb	19	19.4	−0.4	0.00	0.50	
	Mar	28	26.6	1.4	0.30	0.38	
	Apr	26	26.9	−0.9	−0.09	0.54	
	May	22	18.2	3.8	0.99	0.16	
	Jun	20	16.6	3.4	0.86	0.20	
	Jul	24	21.7	2.3	0.52	0.30	
	Aug	31	26.7	4.3	1.00	0.16	
	Sep	72	60.4	11.6	1.64	0.05	type
	Oct	15	13.7	1.3	0.26	0.40	
Intersection evening glare	Feb	46	43.6	2.4	0.46	0.32	
	Mar	28	20.9	7.1	1.60	0.05	type
	Apr	15	14.3	0.7	0.34	0.37	
	May	13	11.1	1.9	0.70	0.24	
	Jun	12	12.1	−0.1	0.04	0.48	
	Jul	26	26.6	−0.6	−0.13	0.55	
	Aug	21	21.8	−0.8	−0.25	0.60	
	Sep	54	46.1	7.9	1.22	0.11	
	Oct	25	26.5	−1.5	−0.27	0.61	
Mid-block morning glare	Feb	12	12.4	−0.4	−0.14	0.56	
	Mar	41	34.1	6.9	1.37	0.09	type
	Apr	29	28.1	0.9	0.28	0.39	
	May	22	19.6	2.4	0.51	0.31	
	Jun	16	14.7	1.3	0.43	0.33	
	Jul	19	22.2	−3.2	−0.84	0.80	
	Aug	25	20.8	4.2	0.98	0.16	
	Sep	61	54.4	6.6	1.03	0.15	
	Oct	3	3.7	−0.7	−0.41	0.66	
Mid-block evening glare	Feb	15	15.0	0	−0.11	0.55	
	Mar	16	18.0	−2	−0.45	0.67	
	Apr	6	6.4	−0.4	0.02	0.49	
	May	13	12.3	0.7	0.23	0.41	
	Jun	6	7.8	−1.8	−0.77	0.78	
	Jul	20	15.1	4.9	1.38	0.08	
	Aug	13	14.8	−1.8	−0.67	0.75	
	Sep	30	28.3	1.7	0.37	0.36	
	Oct	13	12.8	0.2	0.20	0.42	

also conducted to determine the effects of sun glare on different collision types. In this analysis, the collisions along glare-exposed directions in clear weather were used as the case group, referred to as glare collisions, while the collisions occurring on the same direction in overcast weather were set as the control group and referred to as the non-glare related collisions. The null hypothesis was that there was no difference in the portion of certain collision types between glare and non-glare conditions. Under this null hypothesis, the expected collisions of each type can be calculated. For both intersection and mid-block collisions, the data of morning and evening glare periods were combined and analyzed together. Results of this analysis are shown in Table 8. At intersections, it was found that the occurrence of traffic signal violations and stop sign violations were significantly higher (at 95% confidence level) than expected in the glare condition. Also, collisions related to pedestrians and cyclists were also found to be significantly higher (95% confidence level) under the condition of sun glare. On mid-blocks, the results indicate that improper turning and lane changing are significantly increased due to sun glare (95% confidence level). Also, animal-related collisions were significantly higher when drivers were affected by sun glare.

Conclusion and Discussion

This paper has proposed a method to assess the effects of sun glare on road safety within the city of Edmonton. A unique

Table 7. Results of Analysis for the 3 Winter Months

Location type	Glare periods	Existing months	Travel direction	Collisions (in clear weather)		z	Type/antitype	*OR*
				Observed (o)	Expected (e)			
Intersection	Eastbound glare	Jan; Nov	East	27	22	1.22		1.72
			West	19	20	−0.14		0.94
			North	17	20	−0.70		0.72
			South	15	17	−0.47		0.80
	Southbound glare	Jan; Nov; Dec	East	313	322	−0.49		0.94
			West	302	309	−0.46		0.90
			North	318	332	−0.82		0.90
			South	372	342	1.73	type	1.23
	Westbound glare	Jan; Nov	East	20	17	0.64		1.34
			West	22	19	0.99		1.56
			North	13	17	−1.04		0.60
			South	13	15	−0.66		0.72
Mid-block	Southbound glare	Jan; Nov; Dec	North	194	196	−0.14		0.96
			South	132	130	0.16		1.04

*Significant at 95% confidence level.

Table 8. Results of Sun Glare Effects over Different Collision Types

Location type	Collision maneuver type	Glare collisions		z	Type/antitype
		Observed	Expected		
Intersection	Back unsafely	5	7.4	−1.35	
	Fail to yield cyclist	7	4.3	2.05	type*
	Fail to yield pedestrian	22	17.2	2.68	type*
	Fail to yield right of way	3	2.6	0.37	
	Flow too closely	241	250.6	3.51	antitype*
	Improper lane change	29	28.9	0.16	
	Improper turn	9	9.1	−0.1	
	Left of center	3	3.1	0.05	
	Left turn cross path	51	52.7	0.77	
	Ran off road	17	14.9	1.26	
	Signal violation	45	40.5	2.03	type*
	Stop sign violation	46	41.9	1.94	type*
	Strike parked vehicle	4	4.4	0.61	
	Yield sign violation	15	17.3	−1.33	
	Other	3	4.6	−1.18	
Mid-block	Animal action	4	2	1.72	type*
	Back unsafely	6	12.3	3.83	antitype*
	Fail to yield pedestrian	2	1.2	0.22	
	Fail to yield right of way	18	18.7	0.92	
	Flow too closely	132	131.2	0.93	
	Improper lane change	60	54.4	1.98	type*
	Improper turn	7	3.5	2.69	type*
	Left of center	13	11.9	0.53	
	Strike parked vehicle	80	76.7	0.67	
	Run off road	27	28.6	1.27	
	Other	11	10.4	0.24	

*Significant at 95% confidence level.

two-stage safety assessment was proposed, which links the results from modeling sun glare exposure to a thorough glare related collision analysis.

In the sun glare occurrence analysis, the sun glare exposure was modeled for the whole road network by considering factors associated with sun positions, the angle of the glare, the geometry of the roadway alignments, and the terrain configuration. Consequently, the potential sun glare time windows were identified for each month and the corresponding glare-prone locations were plotted using GIS-based visualization maps. The findings of this analysis are summarized as follows:

- The majority of locations affected by morning and evening sun glare consist of the roads with east–west orientation in the grid type road network. Furthermore, during the main winter months of November, December, and January, the sun glare was expected to occur throughout the entire day because of Edmonton's high latitude and continuous low elevation of the sun. However, it was found that during most of the glare times in winter, the sun facing south, meaning that the majority of exposed roads were those with north–south orientation.
- The analysis resulted in the generation of multiple maps that provide a reference of the glare prone locations. These locations were then used to extract glare related collisions and non-glare comparison collisions for the second stage analysis.

In the collision analysis, the collision counts were compared between glare and non-glare conditions to control for other confounding factors. Three research questions were addressed in this analysis and the main findings are summarized as follows:

- The general effects of sun glare were examined by performing the Chi-square test of independence, CFA, and W-test. Similar findings were obtained from all three tests and showed that the sun glare contributed significantly to an increase in collision occurrence, especially at intersections. Quantitatively, the collisions were estimated to increase by approximately 30% during sun glare times.
- The CFA was performed for each month and during the winter period. The results showed that the vision obstruction due to glare during morning and evening periods was especially potent during spring and fall. While during most of the winter glare times, traffic safety in the southbound direction was significantly affected by sun glare.
- The CFA was also performed to check whether sun glare affected various collision types. It was found that there was a higher portion of collisions due to signal or sign violations and failing to yield to pedestrians or cyclists at intersections, while there were higher portions of collisions due to improper turning or lane changing on mid-blocks.

Overall, there are three major contributions in this research. Firstly, the methodology in this paper provides a holistic approach to quantify the effects of sun glare on road safety. By linking sun glare exposure modeling with a thorough glare related collision assessment, this research helps to provide additional insights about the extent to which sun glare affects road safety. Second, this research develops a general knowledge of sun glare occurrence and its modelling. The key factors of this analysis, as it relates to glare angle and glare time-window identification, can be coordinated with current road safety assessments on collision prone locations. If a location is identified as hazardous, a detailed glare analysis might be conducted and, if indeed glare was an issue, provide countermeasures such as placing glare shelters at certain angles or vision assisting/warning signals could be proposed to reduce the effects of glare. Moreover, the findings can also be used to assist in alleviating the glare risks in the planning of future roads from a geometrical perspective. Lastly, the research can also be integrated with intelligent transportation system technologies, such as sun glare route warnings through map navigation, autonomous anti-glare vision machine system, and sun glare driving assist system for the maneuvers that are most likely to be affected.

Future research can analyze the differences that sun glare might have on various road environments. Unfortunately, given the scale of this study, collision related exposure information was not readily available for each road location so future efforts might focus on integrating information on traffic volume, average speeds, and so forth into the analysis. Besides, to account for the absence of detailed DEM, the shadow effects are only addressed by considering the terrain configuration and excluding the downtown area where high-rise buildings dominate the landscape. Such a limitation can be addressed if a more detailed DEM is obtained to generate more precise shadow coverage maps.

Acknowledgments

The authors would like to thank the City of Edmonton and the Office of Traffic Safety for providing the data and for their assistance in facilitating this research.

Author Contributions

The authors confirm contribution to the paper as follows: study conception and design: Danyang Sun, Karim El-Basyouny; data collection: Danyang Sun; analysis and interpretation of results: Danyang Sun, Karim El-Basyouny, Tae J. Kwon; draft manuscript preparation: Danyang Sun, Karim El-Basyouny, Tae J. Kwon. All authors reviewed the results and approved the final version of the manuscript.

References

1. Churchill, A. M., Y. Tripodis, and D. J. Lovell. Sun Glare Impacts on Freeway Congestion: Geometric Model and Empirical Analysis. *Journal of Transportation Engineering*, Vol. 138, No. 10, 2012, pp. 1196–1204.
2. Hagita, K., and K. Mori. The Effect of Sun Glare on Traffic Accidents in Chiba Prefecture, Japan. *Asian Transport Studies*,

Vol. 3, No. 2, 2014, pp. 205–219. https://doi.org/10.11175/eastsats.3.205.
3. Staver, J. Sun Glare Accidents. *Staver Law Group PC*. https://www.chicagolawyer.com/sun-glare-accidents/. Accessed July 6, 2017.
4. Massey, R. *Glare of the Sun Contributes to 3,000 Road Accidents and Is Particularly Dangerous This Time of Year*. http://www.dailymail.co.uk/news/article-2461972/Glare-sun-contributes-3-000-road-accidents-particularly-dangerous-time-year.html. Accessed July 6, 2017.
5. Schreuder, D. A. *Side Lights and Low-Beam Headlights in Built-up Areas*. Institute for Road Safety Research, Voorburg, Netherlands, 1968.
6. McGwin, G., V. Chapman, and C. Owsley. Visual Risk Factors for Driving Difficulty among Older Drivers. *Accident Analysis & Prevention*, Vol. 32, No. 6, 2000, pp. 735–744.
7. Babizhayev, M. A. Glare Disability and Driving Safety. *Ophthalmic Research*, Vol. 35, No. 1, 2003, pp. 19–25.
8. Auffray, B., R. M. Audin, C. M. Monsere, and R. L. Bertini. An Empirical Investigation of the Impacts of Sun-Related Glare on Traffic Flow. Presented at 87th Annual Meeting of the Transportation Research Board, Washington, D.C., 2008.
9. Choi, E.-H., and S. Singh. Statistical Assessment of the Glare Issue-Human and Natural Elements. In *National Center for Statistics and Analysis*, National Highway Traffic Safety Administration, Washington D.C., 2005.
10. Mitra, S. Sun Glare and Road Safety: An Empirical Investigation of Intersection Crashes. *Safety Science*, Vol. 70, 2014, pp. 246–254. https://doi.org/10.1016/j.ssci.2014.06.005.
11. Esri. *Esri Support 10.4*. http://support.esri.com/en/Products/Desktop/arcgis-desktop/arcmap/10-4-1. Accessed July 31, 2017.
12. AltaLIS. *DEM Terrain Products*. http://www.altalis.com/products/terrain/dem.html. Accessed July 31, 2017.
13. Michalsky, J. J. The Astronomical Almanac's Algorithm for Approximate Solar Position (1950–2050). *Solar Energy*, Vol. 40, No. 3, 1988, pp. 227–235. https://doi.org/10.1016/0038-092X(88)90045-X.
14. Environment Canada. *Weather Information*. Environment Canada. https://weather.gc.ca/index_e.html. Accessed June 17, 2017.
15. Jurado-Pina, R., and J. M. Pardillo-Mayora. Methodology to Analyze Sun Glare Impact on Highway under Prolonged Exposure. *Journal of Transportation Engineering*, Vol. 136, No. 12, 2010, pp. 1137–1144.
16. Zwahlen, H. T. Conspicuity of Suprathreshold Reflective Targets in a Driver's Peripheral Visual Field at Night. *Transportation Research Record: Journal of the Transportation Research Board*, 1989. 1213: 35–46.
17. Khumalo, N. *Visibility Improvements through Information Provision Regarding Sun Glare : A Case Study in Cape Town*. Thesis. University of Cape Town, 2014.
18. Eye, A., C. Spiel, and P. K. Wood. Configural Frequency Analysis in Applied Psychological Research. *Applied Psychology*, Vol. 45, No. 4, 1996, pp. 301–327.
19. Siegel, S. *Nonparametric Statistics for the Behavioral Sciences*. McGraw-Hill, New York, 1956.

The Standing Committee on Visibility (AND40) peer-reviewed this paper (18-03717).

Article

TRR
JOURNAL OF THE TRANSPORTATION RESEARCH BOARD

Transportation Research Record
2018, Vol. 2672(16) 93–104

DOI: 10.1177/0361198118774748
journals.sagepub.com/home/trr
SAGE

Investigating the Impact of Fog on Freeway Speed Selection using the SHRP2 Naturalistic Driving Study Data

Md Nasim Khan[1], **Ali Ghasemzadeh**[1], **and Mohamed M. Ahmed**[1]

Abstract

The negative effect of reduced visibility on driver performance has been recognized as one of the main causes of motor vehicle crashes in fog. Although many studies have concentrated on driver behavior during foggy weather in a simulated environment, there is a lack of studies that have addressed the impact of fog on driver behavior and performance in naturalistic settings. This paper utilized the data from the SHRP2 Naturalistic Driving Study (NDS) database to understand driver behavior in general and speed selection in particular during clear and foggy weather conditions. In this study, a comparative preliminary analysis and an ordered logit model were developed to evaluate driver speed behavior in fog and clear weather conditions. Results from the preliminary analysis showed 10% and 3% reduction in speed because of near fog and distant fog, respectively. In addition, results from the speed selection model showed that the odds of reducing speed were 1.31 and 1.28 times higher for drivers traveling in near fog and distant fog, respectively, compared with drivers who were driving in clear weather conditions. However, there is an over-representation of young drivers in the SHRP2 NDS database, which was reflected in the dataset used in this study. Therefore, a more representative sample of age groups might provide different results. The results from this study could provide a better insight into driver speed selection during foggy weather conditions, which can be utilized to improve various safety strategies including variable speed limits.

According to the Federal Highway Administration, roughly 15% of fatal crashes, 19% of injury crashes, and 23% of property-damage-only (PDO) crashes occur in the presence of adverse weather, which results in approximately 5,100 fatal crashes, 304,800 injury crashes, and 922,200 PDO crashes every year (*1*). Many studies have explored the impact of adverse weather on crashes and found that crash rate increases during inclement weather (*2–4*).

Driving in foggy weather is challenging because of reduced visibility, limited contrast, and distorted perception, which causes many accidents every year. From a visual perspective, fog can be described as a reduction in contrast in the visual field (*5*). In fog, drivers face difficulty in perceiving speed and headway, as well as road signs and markings, which are crucial for safe driving (*6*). Fog-related crashes tend to involve multiple vehicles and usually have more fatalities compared with crashes under clear weather conditions (*7*). A previous study showed that foggy conditions may increase crashes specifically in lack of road lighting (*8*). Moore and Cooper found that drivers usually considered the leading vehicle as a means of guidance and drove at a speed similar to the leading vehicle while driving in foggy weather. They stated this tendency to be the main cause of rear-end crashes in foggy weather (*9*).

Yan et al. investigated the effect of foggy weather conditions on driver speed control at different risk levels and found that at the high-risk level, driver speed compensation because of fog did not reduce their crash-involvement risk, though it effectively lowered the crash severity (*10*). A study conducted in a driving simulator environment investigated driver speed perception in foggy weather conditions. They found that drivers perceived their speed as slower than the actual speed in foggy weather conditions (*11*). However, several studies using more sophisticated driving simulators contradict these results. For instance, Owens et al. showed that drivers tended to overestimate their speed and drove at a speed lower than instructed (*12*). In another study, based on 566 surveyed drivers in Florida, Hassan et al. concluded that drivers usually take shorter headways from lead vehicles during limited visibility conditions caused by near fog. The study also mentioned that the variable speed limit (VSL) signs during fog cannot reflect the accurate safe speed limit because of the frequent fluctuation of fog thickness (*13*).

[1]Department of Civil & Architectural Engineering, University of Wyoming, Laramie, WY

Corresponding Author:
Address correspondence to Ali Ghasemzadeh: aghasemz@uwyo.edu

As mentioned in the previous section, the majority of the studies were mainly performed in a simulated environment or utilizing survey questionnaires (*10–15*). However, utilizing real-world data is becoming more and more accessible and popular (*16 –20*).

There is a lack of studies that have examined the impact of fog on driver behavior and performance under naturalistic settings. The data used in this study were collected from the second Strategic Highway Research Program (SHRP2) Naturalistic Driving Study (NDS), which is the largest study on naturalistic driving behavior to date in the U.S. In addition, the Roadway Information Database (RID) was also utilized. The RID was developed mainly to link roadway information to the NDS database.

This study will help in gaining insights into driver dynamics of adapting speed. It will also provide valuable information about how drivers interact with roadway and foggy weather conditions, which can be used for effective countermeasures.

The SHRP2 NDS data have several advantages over studies that are conducted in a simulator or controlled experimental environment. For instance, driver behavior in a simulator or in a controlled environment may not properly represent actual behavior in normal driving conditions. More specifically, driver behavior and performance might be biased, and drivers may not behave naturally, as they do in real life. Naturalistic driving studies can overcome these problems and provide an opportunity to investigate the safety impacts of driver behavior during different weather and roadway conditions.

The overall objective of this study is to assess driver behavior and performance during clear and fog weather conditions by utilizing the NDS and RID database. This will be attained by developing an appropriate NDS and RID data acquisition technique, then conducting a preliminary analysis between clear and foggy weather conditions, and finally developing a speed selection model to identify the major contributing factors that might influence driver speed selection under foggy weather conditions.

Data Source

The SHRP2 NDS data used in this study were requested from the Virginia Tech Transportation Institute (VTTI). The SHRP2 is the largest study on naturalistic driving behavior to date in the U.S. Data were collected from six sites around the United States, including Florida, Indiana, New York, North Carolina, Pennsylvania, and Washington. Participant vehicles were instrumented with a data acquisition system (DAS). The DAS includes forward radar; four video cameras, including one forward-facing color wide-angle camera; accelerometers; vehicle network information; geographic positioning system; onboard computer vision lane tracking, plus other computer vision algorithms; and data storage capability. In addition to the NDS data, the SHRP2 RID was also utilized in this study. The RID database was developed by the Iowa State University Center for Transportation Research and Education. The primary objective of this database is to provide high-quality roadway data that can be linked to the NDS database. The RID contains a comprehensive description of roadway characteristics including curve radius, grade, superelevation, number of lanes, shoulder information, and so forth (*21*).

To effectively identify the NDS trips occurring in fog, weather data from the National Climate Data Center (NCDC) was used. Previous studies have utilized NCDC weather data to identify road weather conditions (*22*, *23*). The NCDC archives weather data from various weather stations nationwide, including radar, satellites, airport weather stations, and military weather stations. Among these data sources, the airport weather stations proved to be the most beneficial to identifying adverse weather events.

Data Reduction

Acquisition and preparation of the NDS data were challenging for various reasons. For instance, identifying the event of interest, processing and reducing video data, and linking the NDS data with the RID and questionnaire data were difficult, labor-intensive, and time-consuming tasks.

To efficiently extract the fog-related trips, a unique method was developed. This method used weather data from the NCDC. In order to identify the spatial-temporal segments of freeways as a potential location for trips occurring in fog, a buffer zone of influence needed to be specified. In this study, a buffer zone of 5 nautical miles (n.m.) was used as a zone of influence. This method was utilized in a previous study and found that the accuracy of identifying fog-related crashes is over 60% (*22*). NDS trips were requested based on the daily weather information to identify all trips affected by adverse weather conditions. By using this process, the research team received data for weather-related trips and matched trips in clear weather conditions. The provided NDS data included video views of forward and backward roadways, basic trip characteristics, and time series data. The received data were then processed and observed manually to filter out the trips that were not related to fog.

To effectively reduce time series and video data an extensive data reduction procedure was conducted. As weather conditions were not consistent throughout the whole trip, each trip was divided into 1-minute segments to ensure consistent weather and traffic conditions within each segment. In addition, video reviewers were provided with observation templates. They extracted environmental and traffic-related data including visibility, surface conditions, freeway segment type, and traffic conditions for each 1-minute segment by observing the videos manually.

Although visibility information could be obtained from the NCDC, the weather station reporting frequency is relatively long, usually 1 hour, and the fluctuation of visibility within 1 hour is common. Moreover, weather stations are mostly mounted at a higher elevation and do not necessarily represent roadway surface visibility. It is worth mentioning that identifying near fog and distant fog based on visibility range from weather stations would not provide real-time visibility conditions. To overcome these limitations, each NDS trip was observed manually and the visibility level was identified for each 1-minute segment.

Considering visibility conditions, foggy weather was classified into two levels including near fog and distant fog. The fog that was close to the NDS vehicle and was more likely to affect driver speed choice and performance was considered as near fog; whereas, fog that was not immediately near to the NDS vehicle but covered the surroundings of the roadway was considered as distant fog. To maintain consistency in reducing video data, video observers were provided with comprehensive training. Video observers reported a weather condition as distant fog if the horizon could not be seen properly but the roadside surroundings (delineators, guardrail, and so forth) can be recognized to some extent, and the information on the road signs and markings can be recognized properly. On the other hand, a weather condition was categorized as near fog if roadside surroundings (delineators, guardrail, and so forth), road markings, and the horizon could not be seen clearly, and the information on road signs is not readable.

The speed limit data, as well as other roadway characteristics such as curve length, radius, superelevation, and so forth, provided in the RID database, were then merged with each 1-minute segment. Finally, driver demographics such as age, income, gender, mileage last year, and so on, collated from the questionnaire survey, were added to each 1-minute segment to create the summary files. These summary files then served as the foundation for modeling and representing driver behavior in foggy weather conditions. It is worth mentioning that online questionnaires were filled out by the NDS participants before involving them in the project. These questionnaires collected information about driver demographics as well as other data about driver sleep, health, and risk-taking behavior (*24*).

Previous studies by the authors using sample trips indicated that the resolution/quality of the forward-facing camera videos during nighttime was very low, especially in adverse weather conditions. Therefore, considering the fact that several variables, including traffic conditions and weather and surface conditions, needed to be reduced from the forward-facing video cameras, nighttime trips were eliminated from the final dataset (*25–27*).

A total of 124 trips in fog and 248 matched trips in clear weather conditions have been fully processed using the described data reduction procedure, which resulted in a total of 7,147 1-minute segments (2,549 in fog, 4,598 in clear weather). As mentioned earlier, this study is trying to provide better insights into driver speed selection under foggy weather conditions, which can lead to providing more realistic speed limits for the VSLs on freeways. Therefore, data were requested for freeways only. However, preliminary analysis showed that some trips have some non-freeway segments at the beginning and at the end of the trips (entering and exiting freeways). Therefore, non-freeway segments were removed from the start and end of these trips. Once the non-freeway segments were removed, the number of 1-minute segments was reduced to 5,587 (i.e., 1,912 in fog and 3,675 in clear weather). A total of 62 drivers between 16 to 79 years of age participated in the selected NDS trips, with more than 80% of drivers aged below 34 years. The number of male and female drivers was almost equal, with a marginally higher percentage (56%) of male drivers. Figure 1 shows the details of the data reduction procedure.

Preliminary Analysis

Summary Statistics

Traffic conditions were categorized into free-flow (Level of service A and B) and mixed traffic condition (Level of service C to F). According to the *Highway Capacity Manual*, free-flow is defined as low-volume roadway conditions, where drivers are free to drive at desired speed and are not constrained by the presence of other vehicles (*28*). In this study, a trip was considered as a free-flow speed when the NDS driver has no leading traffic in any lanes or when a leading vehicle is present at least in one lane, but the NDS driver is still not affected by other vehicles. Other conditions in which NDS drivers were affected by other vehicles were considered as mixed traffic conditions.

Each trip in fog was matched with two trips in clear weather considering the same driver, same vehicle, and same location. Matched trips were requested from the VTTI. As mentioned in a previous study, weather conditions may not be consistent within a trip (*26*). Therefore, considering the possible variations in weather conditions, exact match of the 1-minute segments in fog and clear weather was conducted by importing the longitude and latitude of the trips and eliminating non-matching segments in the ArcGIS software.

Removing the non-matching segments resulted in 5,398 1-minute matching segments (1,867 in fog, 3,531 in clear weather), which was equivalent to nearly 90 hours of driving time and 8,335 kilometers of mileage. The summary statistics of these 5,398 1-minute segments are provided in Table 1.

Speed Distribution

This study investigated the distribution of speeds between clear and foggy weather conditions in various traffic states. From the NDS sample data, it was concluded that the speeds have a Weibull distribution in near fog under free-flow conditions, whereas the speeds were normally distributed in clear weather for the matching dataset. A similar trend was

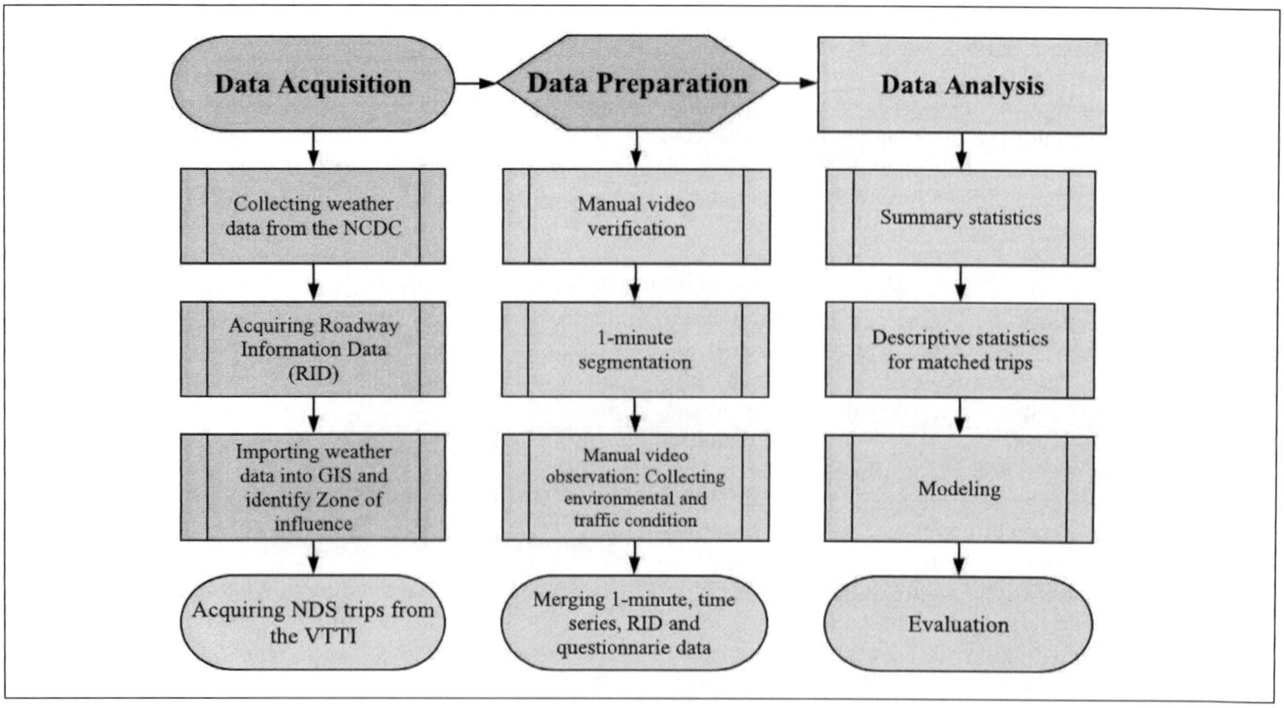

Figure 1. Flow chart summarizing the data reduction and analysis procedure.

Table 1. Summary Statistics of NDS Trips Considered in This Study

	Weather condition	Near fog	Matched clear	Distant fog	Matched clear	Total
Free-flow condition LOS A & B	Number of 1-minute segments	241	539	717	1467	2964
	Total duration (hour)	4.02	8.98	11.95	24.45	49.40
	Total length (km)	418.59	973.72	1260.29	2637.26	5289.86
Congested traffic condition LOS C to F	Number of 1-minute segments	271	405	638	1120	2434
	Total duration (hour)	4.52	6.75	10.63	18.67	40.57
	Total length (km)	340.35	542.85	750.54	1411.11	3044.85
Total number of 1-minute segments		512	944	1355	2587	5398
Total duration (hour)		8.53	15.73	22.58	43.12	89.97
Total length (km)		758.94	1516.57	2010.83	4048.37	8334.71

also found for distant fog, as shown in Figure 2. Speed in free-flow conditions is important for VSL application, as speed selection here is not affected by the interaction with traffic (*25*). Other speed distributions for other scenarios were also examined. Speed distribution during near fog in mixed traffic conditions did not fit a specific distribution. However, speeds during distant fog as well as their matched clear weather conditions in mixed traffic fitted a bimodal distribution, which is common during congestion on freeways (*29*). Figure 2 shows the speed distribution for trips in near fog, distant fog, and matched trips in clear weather under free-flow and congested (i.e., mixed/near traffic) traffic conditions.

Descriptive Analysis

Driver speed behavior, including selection of speeds and accelerations under free-flow conditions, was investigated in clear and foggy weather in order to have a better understanding of driver behavior in different weather conditions. Various statistical tests, including *t*-test, *F*-test, and *Z*-test, were used to compare driver behavior between foggy and clear weather, as shown in Table 2. A *t*-test indicated that the average speed in near fog, as well as in distant fog, was significantly lower than in clear conditions under free-flow traffic. However, speed reduction was greater in near fog compared with distant fog. Speed in near fog and distant

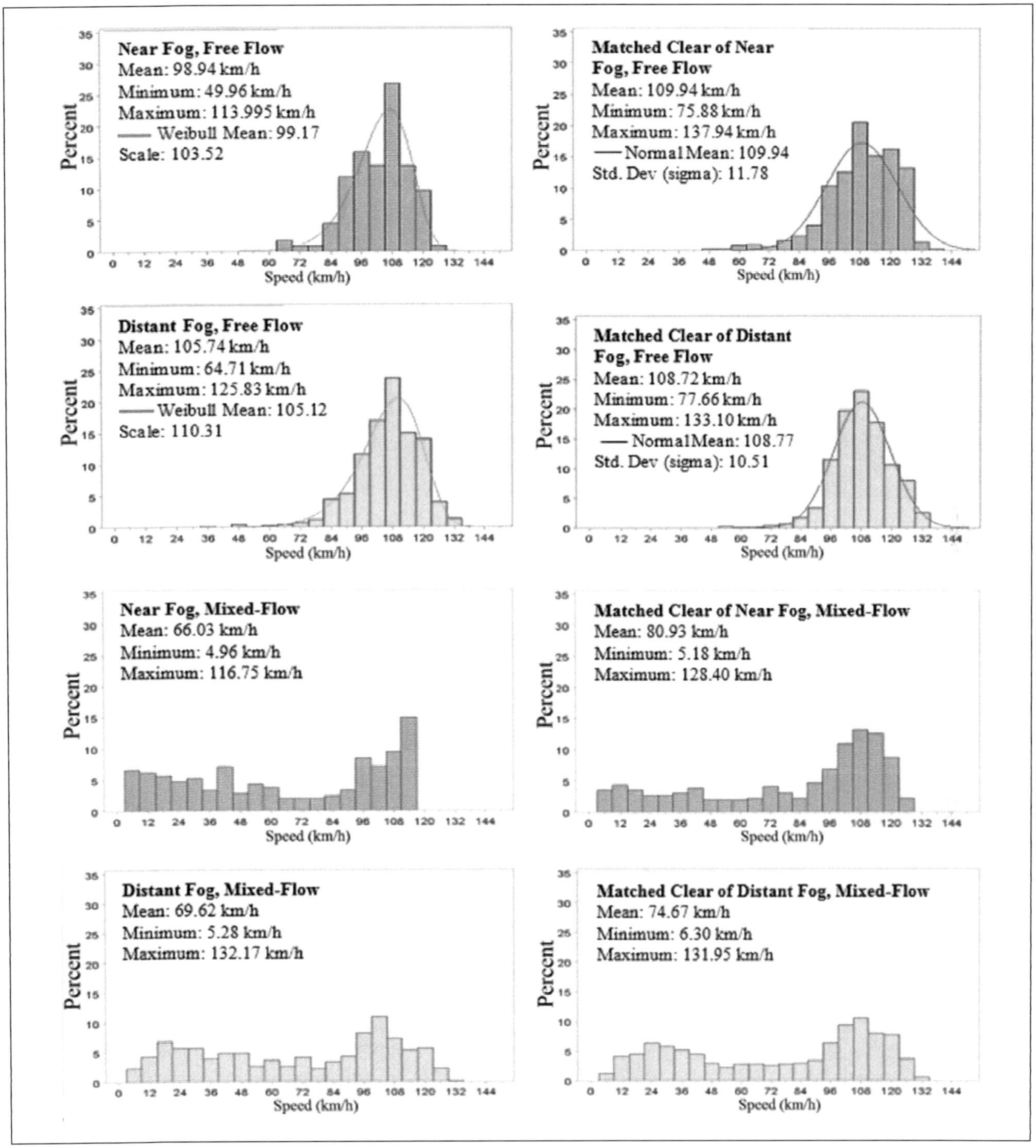

Figure 2. Observed and fitted distributions for speeds during fog and clear weather under free-flow and mixed traffic conditions.

fog was found to be 11 km/h (10% reduction) and 3 km/h (2.8% reduction) lower than the speeds in their matching clear weather conditions, respectively. Previous studies have also concluded similar results of speed reduction. For instance, Liang and Kyte found 6–10 km/h speed reduction as a result of the poor visibility caused by fog (*30*). It was also found that speeds had a higher variability during distant fog compared with clear weather under free-flow traffic, which could be an indication of increased safety risk (*31*, *32*). However, an opposite trend was found for near

Table 2. Descriptive Statistics for NDS Instrumented Vehicles in Fog

		Free-flow traffic							
		Near fog		Matched clear		Distant fog		Matched clear	
	Statistical test	Speed (km/h)	% Speed difference from speed limit	Speed (km/h)	% Speed difference from speed limit	Speed (km/h)	% Speed difference from speed limit	Speed (km/h)	% Speed difference from speed limit
Speed (km/h)	Average	98.944	−1.556	109.944	−8.420	105.739	−4.213	108.722	−7.803
	SD	10.436	10.141	11.785	9.831	11.087	10.070	10.511	8.558
	Min.	49.957	−21.360	75.877	−42.627	64.707	−33.525	77.660	−42.104
	Max.	113.995	43.548	137.942	27.530	125.832	33.385	133.097	26.683
	Median	100.313	−2.754	110.195	−8.100	107.018	−5.005	108.887	−7.847
	t-test	Average speed is significantly lower in near fog. $t(317) = -11.15$, $p < 0.05$ Effect size (Cohen's d) = −0.92				Average speed is significantly lower in distant fog. $t(2009) = -5.84$, $p < 0.05$ Effect size (Cohen's d) = − 0.29			
	F-test	Speed variability is significantly lower in near fog. $F_{523,201} = 1.32$, $p < 0.05$				Speed variability is significantly higher in distant fog. $F_{648,1361} = 1.11$, $p < 0.05$			
	Z-test	Proportion of violation ≥10 km/h above the speed limit is significantly lower in near fog. $Z = -5.73$, $p < 0.05$				No significant difference between the proportion of speeding ≥10 km/h in distant fog and clear weather. $Z = 0.69$, $p > 0.05$.			
Acceleration/ deceleration (g)		Acc. (g)	Dec. (g)	Acc. (g)	Dec. (g)	Acc. (g)	Dec. (g)	Acc. (g)	Dec. (g)
	Average	0.017	−0.013	0.016	−0.012	0.016	−0.019	0.017	−0.016
	SD	0.017	0.014	0.017	0.0140	0.014	0.016	0.013	0.015
	Min.	0.000	−0.061	0.000	−0.067	0.000	−0.070	0.000	−0.067
	Max.	0.082	0.000	0.079	−0.000	0.065	0.000	0.072	0.000
	Median	0.011	−0.009	0.011	−0.006	0.011	−0.012	0.013	−0.011
	t-test	Average acceleration is significantly lower in near fog. $t(453) = 0.13$, $p < 0.05$ Effect size (Cohen's d) = −0.02 No significant difference between the average deceleration in near fog and clear weather. $t(218) = -0.80$, $p > 0.05$ Effect size (Cohen's d) = −0.12				No significant difference between the average acceleration in distant fog and clear weather. $t(1191) = -1.14$, $p > 0.05$ Effect size (Cohen's d) = −0.10 No significant difference between the average deceleration in distant fog and clear weather. $t(816) = -1.60$, $p > 0.05$ Effect size (Cohen's d) = −0.12			
	F-test	No significant difference between the acceleration variability in near fog and clear weather. $F_{1011,362} = 1.09$, $p > 0.05$ No significant difference between the deceleration variability in near fog and clear weather. $F_{86,144} = 1.03$, $p > 0.05$				No significant difference between the acceleration variability in distant fog and clear weather. $F_{384,807} = 1.06$, $p > 0.05$ No significant difference between the deceleration variability in distant fog and clear weather. $F_{263,553} = 1.14$, $p > 0.05$			
	Z-test	No acceleration/ deceleration were found higher/lower than ±0.3 g				No acceleration/ deceleration were found higher/lower than ±0.3 g			

Table 3. Odds Ratio for Speed Behavior

Weather condition	Driving below speed limit	Driving above speed limit	Odds ratio	Confidence interval	Significance level
Near fog	77 (33.7 %)	151 (66.3%)	2.42	1.69 to 3.45	$p < 0.0001$
Matched clear of near fog	386 (17.4%)	1525 (82.6%)			
Distant fog	196 (29%)	480 (71%)	1.90	1.54 to 2.36	$p < 0.0001$
Matched clear of distant fog	253 (17.7%)	1180 (82.3%)			

fog, where speeds in clear weather conditions had more variability.

In addition, the acceleration/deceleration variability was examined, and ±0.3 *g* acceleration/deceleration rates were set as a threshold to identify aggressive braking/acceleration events (*33*). However, no acceleration and deceleration events were found to be higher or lower than ±0.3 *g*, indicating the occurrence of zero aggressive events, as shown in Table 2.

Speed Selection

Speed above the speed limit in fog and respective matching trips in clear weather were examined to determine driver compliance to the speed limit in different weather conditions. It was found that NDS drivers drove consistently above the speed limit in all conditions, including near fog. For instance, NDS drivers drove with a speed 1.6% above the speed limit in near fog; whereas in clear weather NDS drivers drove with a speed 8.4% above the speed limit. Similar results were also found for distant fog. A *Z*-test as shown in Table 2 indicates that the violation of speed greater than 10 km/h was significantly lower in near fog compared with matching trips in clear weather. However, no significant difference was found between the violation of speed greater than 10 km/h in distant fog and corresponding matched clear trips, indicating no effect of distant fog on speeding behavior.

According to Table 3, the majority of the drivers drove with a speed above the limit. For instance, speeds of about 66% of the trips in near fog and almost 83% of the trips in matching clear weather were above the speed limit. Similarly, about 71% of the trips in distant fog and 82% of the trips in respective clear weather were driven at speeds more than the speed limit. Table 3 also indicates that speed reduction was more likely to occur in foggy weather conditions in comparison with the matched trips in clear weather conditions. The odds ratios of driving below the speed limit, in general, were 2.4 and 1.9 times more likely to be in near fog and distant fog, respectively, than matching trips in clear weather conditions.

Modeling Speed Selection: Ordered Logit Model

To better understand the factors affecting driver speed selection in different weather conditions an ordered logit model was developed. The model was calibrated utilizing a dataset of 5,587 1-minute segments occurring in various weather and traffic conditions. A logit model has various advantages over other models. For instance, predictors in the logit model can be continuous, categorical, or a mixture of both. In addition, independent variables do not have to be normally distributed or have equal variance in each group (*34*). Table 4 shows the summary of different variables used in the speed selection model. The response variable of the model is speed selection, which is classified into four levels, based on the median percent speed reduction above or below the speed limit $\frac{(Speed\ limit - speed)}{Speed\ limit}$. The four-quartile intervals were defined as: (1) More than 10% increase in speed, (2) 0–10% increase in speed, (3) 0–28% reduction in speed, and (4) more than 28% reduction in speed. These intervals were selected based on quartile values in order to ensure sufficient sample size in each category. The remaining variables are explanatory variables including environmental variables, traffic conditions, driver demographics, and roadway factors.

Speed Selection Model Results

The log likelihood ratio (LR) was used to confirm the fitness of the model. The LR test statistic as shown in Table 5 falls into the rejection region with a *p*-value <0.05, which indicates the overall explanatory variables of the model have significant effect on the response at a statically significant level of 95%. To check the possible presence of multicollinearity, the variance inflation factor (VIF) was calculated for each predictor. The VIF measures how much the variance of an estimated regression coefficient increases if predictors are correlated. A VIF between 5 and 10 shows a high correlation between predictors and a VIF greater than 10 indicates that the regression coefficients are poorly estimated as a result of multicollinearity (*35*). However, the VIF value of all the predictors in the speed selection model fell below 2.5, indicating no multicollinearity problem. Only the statistically significant variables were retained in the final model. Table 5 shows the results of the speed selection model.

Discussion of Key Factors

Fourteen variables and three interaction terms were found to be significant in the speed selection model. As expected, fog

Table 4. Data Descriptions for Speed Selection Model

Variable	Description	Type	Levels
Response variable			
Speed selection	Percent speed reduction above or below the speed limit	Ordinal	1 = More than 10% increase in speed 2 = 0–10% increase in speed 3 = 0–28% reduction in speed 4 = More than 28% reduction in speed
Explanatory variables			
Environmental variables			
Weather conditions	Predominant weather conditions in 1-min video observation	Categorical	1 = Clear 2 = Distant fog 3 = Near fog
Visibility	Predominant visibility conditions in 1-min video observation	Categorical	1 = Not affected 2 = Affected
Surface conditions	Predominant surface conditions in 1-min video observation	Binary	1 = Dry 2 = Wet
Traffic variables			
Traffic condition	Predominant traffic conditions in 1-min video observation	Binary	1 = Free flow 2 = Mixed flow
Speed limit	Predominant speed limit conditions in 1-min segment	Categorical	1 =<55 2 = 55–60 3 = 65–70
Driver demographics			
Gender	The gender the participant	Binary	1 = Male 2 = Female
Age	The age group corresponding to the driver's birthdate	Categorical	1 = Less than 40 years 2 = Greater than 40 years
Education	The highest completed level of education of the participant	Categorical	1 = High school diploma or G.E.D. 2 = Some education beyond high school but no degree and college degree 3 = Some graduate or professional school, but no advanced degree (e.g., J.D.S., M.S. or Ph.D.) and advanced degree (e.g., J.D.S., M.S. or Ph.D.)
Marital status	The participant's marital status	Categorical	1 = Single 2 = Married 3 = Other (divorced, widow, unmarried partners)
Driver mileage last year	The approximate number of miles the participant drove last year	Categorical	1 = Less than 10,000 miles 2 = Between 10,000 to 20,000 miles 3 = Greater than 20,000 miles
Driving experience	Number of years of driving experience	Categorical	1 = Less than 10 years 2 = Greater than 10 years
Roadway factors			
Bridge	Presence of bridge	Binary	1 = No bridge 2 = On bridge
Curve	Presence of curve	Binary	1 = No curve 2 = On curve
Superelevation	Average superelevation of the road in 1-min segment	Continuous	–
Curve length	Average length of curve of the road in 1-min segment	Continuous	–

had a significant effect on speed selection. Results showed that drivers were likely to travel at significantly lower speeds during foggy weather conditions; more specifically, the odds of drivers reducing their speeds were 1.31 and 1.28 times higher for drivers traveling in near fog and distant fog, respectively, in comparison with drivers who were driving in

Table 5. Estimation of Ordered Logit Model for Speed Selection

Parameter		DF	Estimate	Standard error	Wald chi-square	p-value	Odds ratio	95% confidence interval	
Intercept	4	1	−1.4876	0.2516	34.9552	<.0001	0.226	–	–
Intercept	3	1	0.0017	0.2508	0.0000	0.9945	1.002	–	–
Intercept	2	1	1.4051	0.2509	31.3594	<.0001	4.076	–	–
Weather	Clear	1	–	–	–	–	–	–	–
	Distant fog	1	0.2528	0.0741	11.6346	0.0006	1.288	1.113	1.490
	Near fog	1	0.2726	0.1234	4.8808	0.0272	1.313	1.031	1.673
Visibility	Not affected	1	–	–	–	–	–	–	–
	Affected	1	0.4953	0.1377	12.9452	0.0003	1.641	1.253	2.149
Surface condition	Dry	1	–	–	–	–	–	–	–
	Wet	1	0.7642	0.1878	16.5508	<.0001	2.147	1.486	3.103
Traffic condition	Free flow (A–B)	1	–	–	–	–	–	–	–
	Mixed flow (C–F)	1	1.9217	0.0585	1080.4932	<.0001	6.833	6.093	7.662
Speed limit	<55 mph	1	–	–	–	–	–	–	–
	55–60 mph	1	−1.4557	0.2215	43.1710	<.0001	0.233	0.151	0.360
	65–70 mph	1	−1.8661	0.2238	69.5241	<.0001	0.155	0.100	0.240
Gender	Male	1	–	–	–	–	–	–	–
	Female	1	−0.2529	0.0580	18.9851	<.0001	0.777	0.693	0.87
Age	<40 years	1	–	–	–	–	–	–	–
	>40 years	1	0.1628	0.0801	4.1344	0.0420	1.177	1.006	1.377
Education	High school	1	–	–	–	–	–	–	–
	Beyond high school	1	0.3776	0.1107	11.6391	0.0006	1.459	1.174	1.812
	Advance degree	1	0.9529	0.1267	56.5368	<.0001	2.593	2.023	3.324
Marital status	Single	1	–	–	–	–	–	–	–
	Married	1	0.2804	0.0794	12.4663	0.0004	1.324	1.133	1.547
	Others	1	−0.3879	0.0936	17.1765	<.0001	0.678	0.437	0.741
Driver's mileage last year	<10000 miles	1	–	–	–	–	–	–	–
	10,000 – 20,000 miles	1	−0.5999	0.0731	67.4055	<.0001	0.549	0.475	0.633
	>20,000 miles	1	−0.3879	0.0936	17.1765	<.0001	0.678	0.565	0.815
Driver's experience	<10 years	1	–	–	–	–	–	–	–
	>10 years	1	0.5890	0.0934	39.7381	<.0001	1.802	1.500	2.165
Bridge	No	1	–	–	–	–	–	–	–
	Yes	1	0.7457	0.2241	11.0670	0.0009	2.108	1.358	3.27
Super elevation	–	1	0.0258	0.0113	5.2036	0.0225	1.026	1.004	1.049
Curve length	–	1	− 0.0011	0.0002	21.4193	<.0001	0.999	0.998	1.000
Speed limit × curve length	55–60 mph, curve length	1	0.0010	0.0002	15.8866	<.0001	1.001 at speed limit 55–60 mph	1.000	1.002
	65–70 mph, curve length	1	0.0011	0.0002	20.9590	<.0001	1.001 at speed limit 65–70 mph	1.000	1.002
Driver's experience × visibility	>10 years, affected	1	−0.6644	0.1646	16.2971	<.0001	0.515 at affected visibility	0.373	0.710
Curve × weather	Curve × near fog		0.3835	0.1897	4.0838	0.0433	1.467 at near fog	1.012	2.128

Fit statistics:
Likelihood ratio test: $\chi^2 = 2284.36$, df = 24, p-value <0.0001.
Score test for the proportional odds assumption: $\chi^2 = 1313.20$, df = 48, p-value <0.0001.
Akaike information criterion (AIC) = 13254.61.
−2 log L = 13200.614.

clear weather conditions. Driving over the speed limit could be hazardous especially during inclement weather conditions including fog, because drivers might not have enough time to respond to or mitigate an unexpected event (*36*). This study showed that drivers reduced their speeds to compensate for the negative effect of fog on the primary driving tasks.

Findings related to visibility indicated that the odds of drivers reducing their speeds were 1.64 times greater for drivers who were driving in affected visibility versus those driving in good visibility conditions. A similar result was also found for surface conditions. Wet surface was found to have a significant impact on speed reduction. More clearly, the odds of drivers reducing speeds on wet surfaces were 2.15 times higher compared with dry surfaces.

Traffic conditions had a positive coefficient as expected. Controlling for all other variables, drivers were 6.83 times more likely to reduce their speeds in mixed traffic conditions (level of service C to F) compared with free-flow conditions (level of service A and B). It was found that female drivers were less likely to reduce their speed compared with male drivers. More clearly, the odds of female drivers reducing their speeds were 1.29 times less compared with male drivers (OR = 0.777). As expected, older drivers had more speed reduction compared with young drivers. More specifically, the odds of having more speed reduction percentage were 1.18 times higher for drivers older than 40 years compared with drivers of 40 years of age or younger. Education level also came out to be significant in the model. It was found that with the increase of education level, drivers became more compliant to the speed limit. For instance, the odds of a driver with an advanced degree were 2.59 times more likely to reduce speed compared with a driver who is a high school graduate. Marital status was also found to be significant, with a usual trend of married drivers being the safest compared with single drivers (*37*). More clearly, married drivers were 1.32 times more likely to reduce speed compared with single drivers.

Several factors related to the roadway, including the presence of bridge, superelevation, and curve length were found to have a significant effect on driver speed selection. Considering interaction terms, it was found that at an affected visibility condition, experienced drivers (driving experience >10 years) were 48% less likely to reduce speed compared with less experienced drivers (driving experience <10 years), which indicates experienced drivers are usually more confident in reduced visibility compared with less experienced drivers (*38*). Similarly, the interaction between weather condition and curve indicated that the drivers were 1.47 times more likely to reduce their speed on curves compared with their speed on tangents during near fog.

Conclusions

The main focus of this study was to attain better insights into driver behavior in general and speed selection in particular during clear and foggy weather conditions using the SHRP2 NDS dataset. The preliminary analysis showed a Weibull speed distribution in near fog under free-flow conditions, whereas the speeds were normally distributed in clear weather for the matching dataset (i.e., same vehicle, driver, route, and traffic state). Descriptive analysis indicated about 10% reduction in speed during near fog and about 3% reduction in speed during distant fog. The results from the ordered logit model revealed that weather-related factors including the presence of fog, visibility, and surface conditions have a significant impact on driver speed selection behavior. For instance, results showed that drivers were more likely to select significantly lower speeds during foggy weather conditions. More specifically, the odds of drivers reducing their speeds from the posted speed limit were 1.31 and 1.28 times higher for drivers traveling in near fog and distant fog, respectively, compared with drivers who were driving in clear weather conditions. As mentioned before, the majority of the participants in the SHRP2 NDS were young (39% of the NDS drivers were below 25 years old). Considering the fact that the data used in this study represent the age distribution of the actual SHRP2 NDS data, a more representative sample of age groups might provide different results.

Foggy weather conditions can negatively affect driver speed perception and ability to see objects on the roadway, which is one of the main causes of rear-end and lane departure crashes on freeways. Advanced driver assistance systems (ADAS) such as adaptive cruise control (ACC), collision avoidance systems, collison warning system, dynamic brake support, autonomous emergency braking, lane departure warning, and so forth, are currently being used to improve roadway safety in different adverse weather conditions including fog. The main focus of these technologies is to prevent crashes by detecting a conflict, alerting drivers, and aiding in taking appropriate and timely actions. For instance, the ACC can assist drivers to keep a safe distance from a lead vehicle while maintaining a chosen speed (*39*). However, most of the ADAS are based on machine vision techniques, which might be inefficient during adverse weather conditions because of the difficulty of detecting objects in adverse weather. In comparison, sensor-based technologies are usually more effective as they are less susceptible to adverse weather. As the ACC works based on ultrasonic, laser, or LiDAR sensors, it is more effective in adverse weather conditions including fog compared with other machine-based ADAS.

Evaluating driver behavior and performance under the influence of reduced visibility caused by foggy weather conditions is extremely important for developing safe driving strategies, including VSL. Many roadways across the U.S. currently have weather-based VSL systems to ensure safe driving environments during adverse weather. Current VSL systems mainly collect traffic information from external sources, including inductive loop detector, overhead radar, and closed-circuit television. However, human factors, especially driver behavior and performance such as selection of speed and acceleration during adverse weather, are neglected because of the lack of appropriate driver data. The SHRP2 NDS database has huge potential in becoming a good source for driver data. The findings from this study indicated that the NDS data could be effectively utilized to identify trips in foggy weather conditions and to assess the impacts of fog on driver behavior and performance.

The results from this study provided insights into incorporating naturalistic speed selection behavior in VSL systems. While the vast majority of VSL systems are based on road weather iformation system data, previous studies noted many limitations of these systems. Utilization of 1-minute real-time weather and surface conditions and visibility limits may improve VSL logics significantly. With the evolution of connected vehicles, machine vision, and other real-time weather social networks such as WeatherCloud, more accurate real-time data similar to the NDS data will be available in the near future. This study provided early insights into using similar data collected from NDS.

Acknowledgments

This work was conducted under the second Strategic Highway Research Program (SHRP2), which is administrated by the Transportation Research Board of the National Academies, and it was sponsored by the Federal Highway Administration in cooperation with the American Association of State Highway and Transportation Officials (AASHTO).

Author Contributions

The authors confirm contribution to the paper as follows: study conception and design: Ali Ghasemzadeh, Mohamed Ahmed; data preparation and reduction: Ali Ghasemzadeh, Nasim Khan; analysis and interpretation of results: Nasim Khan, Ali Ghasemzadeh, Mohamed Ahmed; draft manuscript preparation: Nasim Khan, Ali Ghasemzadeh, Mohamed Ahmed. All authors reviewed the results and approved the final version of the manuscript.

References

1. *How Do Weather Events Impact Roads?* FHWA Road Weather Management. https://ops.fhwa.dot.gov/weather/q1_roadimpact.htm. Accessed July 5, 2017.
2. Fartash, H., M. Hadi, and Y. Xiao. Utilization the HCM Urban Facility Procedures for the Estimation and Real-Time Prediction of Travel Time with Consideration of Rain Impacts. Presented at 95th Annual Meeting of the Transportation Research Board, Washington, D.C., 2016.
3. Ghasemzadeh, A., and M. M. Ahmed. A Probit-Decision Tree Approach to Analyze Effects of Adverse Weather Conditions on Work Zone Crash Severity Using Second Strategic Highway Research Program Roadway Information Dataset. Presented at 96th Annual Meeting of the Transportation Research Board, Washington, D.C., 2017.
4. Fartash, H., M. Hadi, Y. Xiao, T. Wang, and R. Ponnaluri. Assessing Need for Systemwide Ramp Metering Installation Warrants. *Transportation Research Record: Journal of the Transportation Research Board*, 2017. 2616: 27–38.
5. Brooks, J. O., M. C. Crisler, N. Klein, R. Goodenough, R. W. Beeco, C. Guirl, P. J. Tyler, A. Hilpert, Y. Miller, and J. Grygier. Speed Choice and Driving Performance in Simulated Foggy Conditions. *Accident Analysis & Prevention*, Vol. 43, No. 3, 2011, pp. 698–705.
6. Hamilton, B. A. Ten-Year Averages from 2005 to 2014. Retrieved from Road Weather Management Program - NHTSA: http://www.ops.fhwa.dot.gov/weather/q1_roadimpact.htm. Accessed July 5, 2017.
7. Whiffen, B., P. Delannoy, and S. Siok. *Fog: Impact on Road Transportation and Mitigation Options*. National Highway Visibility Conference, Madison, Wisconsin, USA. 2004, pp. 1–12.
8. Wanvik, P. O. Effects of Road Lighting: An Analysis Based on Dutch Accident Statistics 1987–2006. *Accident Analysis & Prevention*, Vol. 41, No. 1, 2009, pp. 123–128.
9. Moore, R. L., and L. Cooper. *Fog and Road Traffic*. Report No. TRRL-LR-446, Transport and Road Research Laboratory, Crowthorne, Berkshire, UK, 1972.
10. Yan, X., X. Li, Y. Liu, and J. Zhao. Effects of Foggy Conditions on Drivers' Speed Control Behaviors at Different Risk Levels. *Safety Science*, Vol. 68, 2014, pp. 275–287.
11. Snowden, R. J., N. Stimpson, and R. A. Ruddle. Speed Perception Fogs Up as Visibility Drops. *Nature*, Vol. 392, No. 6675, 1998, p. 450.
12. Owens, J. M., L. Angell, J. M. Hankey, J. Foley, and K. Ebe. Creation of the Naturalistic Engagement in Secondary Tasks (NEST) Distracted Driving Dataset. *Journal of Safety Research*, Vol. 54, 2015, pp. 33–e29.
13. Hassan, H. M., M. A. Abdel-Aty, K. Choi, and S. A. Algadhi. Driver Behavior and Preferences for Changeable Message Signs and Variable Speed Limits in Reduced Visibility Conditions. *Journal of Intelligent Transportation Systems*, Vol. 16, No. 3, 2012, pp. 132–146.
14. Kamrani, M., S. M. H. E. Abadi, and S. R. Golroudbary. Traffic Simulation of Two Adjacent Unsignalized T-junctions During Rush Hours Using Arena Software. *Simulation Modelling Practice and Theory*, Vol. 49, 2014, pp. 167–179.
15. Mehrara Molan, A., and A. Abdi Kordani. Multi-Body Simulation Modeling of Vehicle Skidding and Roll over for Horizontal Curves on Longitudinal Grades. Presented at 93rd Annual Meeting of the Transportation Research Board, Washington, D.C., 2014.
16. Kamrani, M., B. Wali, and A. J. Khattak. Can Data Generated by Connected Vehicles Enhance Safety? Proactive Approach to Intersection Safety Management. *Transportation Research Record: Journal of the Transportation Research Board*, 2017. 2659: 80–90.
17. Ghasemzadeh, A., Hammit, B.E., Ahmed, M.M., Young, R.K. Parametric Ordinal Logistic Regression and Non-parametric Decision Tree Approaches for Assessing the Impact of Weather Conditions on Driver Speed Selection Using Naturalistic Driving Data. *Transportation Research Record: Journal of the Transportation Research Board*, 2018. DOI: 10.1177/0361198118758035
18. Ghasemzadeh, A., and M. A. Mohamed. Utilizing Naturalistic Driving Data for In-depth Analysis of Driver Lane-keeping

Behavior in Rain: Non-parametric Mars and Parametric Logistic Regression Modeling Approaches. *Transportation Research Part C: Emerging Technologies*, Vol. 90, 2018, pp. 379–392. https://doi.org/10.1016/j.trc.2018.03.018

19. Ahmed, M. M., and A. Ghasemzadeh. The Impacts of Heavy Rain on Speed and Headway Behaviors: An Investigation Using the shrp2 Naturalistic Driving Study Data. *Transportation Research Part C: Emerging Technologies*, Vol. 91, 2018, pp. 371–384. https://doi.org/10.1016/j.trc.2018.04.012
20. Ghiasi, A., J. Ma, F. Zhou, and X. Li. Speed Harmonization Algorithm Using Connected Autonomous Vehicles. Presented at 96th Annual Meeting of the Transportation Research Board, Washington, D.C., 2017.
21. McLaughlin, S. B., and J. M. Hankey. *SHRP 2 Report S2-S31-RW-3: Naturalistic Driving Study: Linking the Study Data to the Roadway Information Database*. Transportation Research Board of the National Academies, Washington, D.C., 2015.
22. Ahmed, M. M., M. Abdel-Aty, J. Lee, and R. Yu. Real-time Assessment of Fog-related Crashes Using Airport Weather Data: A Feasibility Analysis. *Accident Analysis and Prevention*, Vol. 72, 2014, pp. 309–317.
23. Hassan, H. M., and M. A. Abdel- Aty. Predicting Reduced Visibility Related Crashes on Freeways Using Real-time Traffic Flow Data. *Journal of Safety Research*, Vol. 45, 2013, pp. 29–36.
24. Dingus, T. A., and J. M. Hankey. *SHRP 2 Report S2-S06-RW-1: Naturalistic Driving Study: Technical Coordination and Quality Control*. Transportation Research Board, Washington, D.C., 2015.
25. Ahmed, M. M., and A. Ghasemzadeh. Exploring the Impacts of Adverse Weather Conditions on Speed and Headway Behaviors Using the SHRP2 Naturalistic Driving Study Data. Presented at 96th Annual Meeting of the Transportation Research Board, Washington, D.C., 2017.
26. Ahmed, M. M., A. Ghasemzadeh, Eldeeb, S. H. Gaweesh, J. Clapp, K. Ksaibati, and R. Young. *Driver Performance and Behavior in Adverse Weather Conditions: An Investigation Using the SHRP2 Naturalistic Driving Study Data—Phase 1*. No. FHWA-WY-16/08F, U.S. Department of Transportation, Washington, D.C., 2015. Record URL: https://rosap.ntl.bts.gov/view/dot/30814
27. Ghasemzadeh, A., and M. M. Ahmed. Drivers' Lane-Keeping Ability in Heavy Rain: Preliminary Investigation Using SHRP 2 Naturalistic Driving Study Data. *Transportation Research Record: Journal of the Transportation Research Board*, 2017. 2663: 99–108.
28. *Highway Capacity Manual: A Guide for Multimodal Mobility Analysis,* 6th ed. Transportation Research Board, Washington, D.C., 2016.
29. Jun, J. Understanding the Variability of Speed Distributions Under Mixed Traffic Conditions Caused by Holiday Traffic. *Transportation Research Part C: Emerging Technologies*, Vol. 18, No. 4, 2010, pp. 599–610.
30. Liang, W. L., and M. Kyte. Effect of Enviromental Factors on Driver Speed. *Transportation Research Record: Journal of the Transportation Research Board*, 1998. 1635: 155–161.
31. Abdel-Aty, M. A., H. M. Hassan, M. Ahmed, and A. S. Al-Ghamdi. Real-time Prediction of Visibility Related Crashes. *Transportation Research Part C: Emerging Technologies*, Vol. 24, 2012, pp. 288–298.
32. Ahmed, M., and M. Abdel-Aty. A Data Fusion Framework for Real-time Risk Assessment on Freeways. *Transportation Research Part C: Emerging Technologies*, Vol. 26, 2013, pp. 203–213.
33. Bagdadi, O. Assessing Safety Critical Braking Events in Naturalistic Driving Studies. *Transportation Research Part F: Traffic Psychology and Behaviour*, Vol. 16, 2013, pp. 117–126.
34. Jaeger, T. F. Categorical Data Analysis: Away from ANOVAs (transformation or not) and Towards Logit Mixed Models. *Journal of Memory and Language*, Vol. 59, No. 4, 2008, pp. 434–446.
35. Marquardt, D. W. Generalized Inverse, Ridge Regression, Biased Linear Estimation, and Nonlinear Estimation. *Technometrics*, Vol. 12, No. 1, 1970, pp. 591–612.
36. Ghasemzadeh, A., and M. M. Ahmed. Crash Characteristics and Injury Severity at Work Zones Considering Adverse Weather Conditions in Washington Using SHRP 2 Roadway Information Database. Presented at 95th Annual Meeting of the Transportation Research Board, Washington, D.C., 2016.
37. Tazul Islam, M., L. Thue, and J. Grekul. Understanding Traffic Safety Culture: Implications for Increasing Traffic Safety. *Transportation Research Record: Journal of the Transportation Research Board*, 2017. 2635: 79–89.
38. Campbell, K. L. The SHRP 2 Naturalistic Driving Study: Addressing Driver Performance and Behavior in Traffic Safety. *TR News*, No. 282, September–October 2012. pp. 30–35.
39. Horrey, W. J., A. L. Alexander, and C. D. Wickens. Does Workload Modulate the Effects of In-vehicle Display Location on Concurrent Driving and Side Task Performance. *Presented at Driving Simulation Conference North America*, Dearborn, Michigan, 2003, pp. 1–20.

The Standing Committee on Visibility (AND40) peer-reviewed this paper (18-06286).

Article

TRR
JOURNAL OF THE TRANSPORTATION RESEARCH BOARD

Transportation Research Record
2018, Vol. 2672(16) 105–112
© National Academy of Sciences: Transportation Research Board 2018
Article reuse guidelines:
sagepub.com/journals-permissions
DOI: 10.1177/0361198118782762
journals.sagepub.com/home/trr
SAGE

Effects of Mounting Height, Offset Distance, and Number of Light Towers on Drivers' Visual Performance and Discomfort Glare in Work Zones

Rajaram Bhagavathula[1] and Ronald B. Gibbons[1]

Abstract

Portable light towers are a significant source of glare to motorists entering a work zone. Although existing research has evaluated the effect of light tower orientation on visibility and glare, the effects of factors like mounting height, offset distance from the roadway, and number of light towers in the work zone, on visual performance and discomfort glare is not known. Understanding these relationships can help in developing illuminating guidelines for work zones that can reduce glare for drivers. The goal of this paper is to understand the effect of mounting height, offset distance to the roadway, and number of light towers in the work zone on drivers' visual performance and discomfort glare. Participants drove through a realistic work zone and evaluated portable light towers in varying mounting heights, offset distances, and number of light towers in the work zone. Results showed that the mounting height and offset distances play a critical role in affecting the driver's visual performance and discomfort glare rating. Portable light towers, irrespective of wattage and lumen output, at lower than a mounting height of 20 ft and closer to the roadway result in decreasing driver visual performance and increasing their discomfort glare. Portable light towers should be mounted at a height of at least 20 ft and balloon light towers with higher wattage (4,000 W and greater) and lumen output (400,000 lumens and greater) should be located at an offset distance of at least 10 ft from the roadway.

Portable light towers are commonly used in nighttime work zones to provide illumination to workers. However, they are a significant source of glare to motorists entering the work zone. Glare in the eyes of the drivers entering the work zone could reduce their visibility and potentially increase the risk of a crash. Great care should be taken when prescribing lighting for work zones so that workers have adequate light levels to effectively and safely complete their tasks without introducing glare for drivers entering the work zone.

Prior research in this area examined the effect of several commercially available portable light towers and their orientation on drivers' visual performance and their perceptions of visibility and glare (*1*). In the above-mentioned study, three commercially available portable light towers (4000-W metal halide, LED and balloon type) in three orientations (angle to the drivers' line of sight: 45, 90 and 135 degrees) were used. Visual performance and perception of visibility and glare were evaluated in a realistic work zone designed to simulate a work zone on a limited access highway for the portable light tower types in all the three orientations. The results of this research showed that portable light towers aimed away from or perpendicular to drivers' line of sight (90 degrees or greater) resulted in increasing the driver's visual performance (longer detection distances) and lowering their perception of glare (lower glare ratings). However, the effect of mounting height and the location of the portable light towers with respect to the roadway on visual performance were not reported. Further, newer portable light sources like balloon light towers have been increasingly used in work zones. These newer light sources cannot be aimed and therefore care must be taken to determine an ideal location for these light towers so that they do not negatively affect the visibility of the drivers who are entering the work zone. Previous studies (*1*, *2*) have shown that improper aiming could limit the visibility of objects in the work zones. Thus, it is important to understand the effect of mounting height, location of the light towers with respect to the roadway (or offset distance) on visual performance, and discomfort glare of drivers. Understanding this relationship can help in recommending appropriate illumination guidelines for nighttime work zones, that can result in increased safety for motorists and workers.

This paper has two goals. The first is to evaluate objectively the effects of mounting heights, offset distances, and

[1]Virginia Tech Transportation Institute, Blacksburg, VA

Corresponding Author:
Address correspondence to Rajaram Bhagavathula:
Rbhagavathula@vtti.vt.edu

Table 1. List of Independent Variables and Their Categorical Values

Independent variables	Levels
Age	Older (60+ years) Younger (18–35 years)
Light tower type	4000-W metal halide 4000-W balloon 800-W balloon LED
Mounting height (only 4000-W metal halide and 4000-W balloon)	15 ft 20 ft 30 ft
Offset distance (only 4000-W balloon)	0 ft (in the lane) 10 ft (in the shoulder) 20 ft (off the shoulder)
Number of luminaires (only 800-W balloon LED)	1 2

number of light towers on driver visual performance, especially on limited-access highways. The second is to understand the effects of the above-mentioned variables on the discomfort glare of drivers. Results from this study will add to the body of knowledge on work zone lighting, and glare. These results could also help in developing lighting specifications for work zones.

Methods

Participants

Twenty-four participants were recruited to participate as drivers in this study. The participant sample was divided into two distinct age groups. The first group consisted of younger drivers whose ages ranged from 18 to 35 years (mean *[M]* = 25.4 years, standard deviation *[SD]* = 3.8 years). The second group consisted of older drivers who were 60 years and older (*M* = 63.5 years, *SD* = 2.7 years). Both age groups comprised a participant sample with diverse visual capabilities and driving experiences. All the participants had a valid drivers' license and a minimum visual acuity of 20/40 (corrected).

Experimental Design

A repeated measures experimental design was used to evaluate the effects of mounting height, offset distance, and number of light towers on visual performance and discomfort glare. Visual performance was assessed by measuring detection and recognition distances of a worker as the participants drove through a simulated work zone on the Virginia Smart Road. Discomfort glare was assessed by means of a rating scale. The simulated work zone was set up in such a way that the lane closure was on the right lane when the participants were traveling in one direction and in the left lane when the participants were traveling in the opposite direction. Having the participants travel in both directions saved time and required fewer runs to collect the required data.

The independent variables and their levels are summarized in Table 1. In each experimental session, participants encountered all conditions. The presentation of the mounting heights, offset distances, and the number of light towers were counterbalanced to minimize order effects. Presentation of the simulated worker was also randomized with blanks (no work presentation) to actively discourage participants from guessing.

Independent Variables

Types of Portable Light Towers. Three types of commercially available portable light towers were used (Figure 1). The first was a conventional metal halide portable light tower with four 1,000-W (440,000 lumens) metal halide luminaires. The second was a balloon light tower (Manufacturer: 812 Illumination, model 4000-W HID) with four 1,000-W metal halide luminaires (440,000 lumens) enclosed within a balloon, which diffused the light. The third was also a smaller balloon light tower with an 800-W LED luminaire (Manufacturer: 812 Illumination; 84,000 lumens). To account for the different wattages and lumen output of the portable light tower types, the vertical illuminance on the simulated worker was matched at 50 lux across all the light tower types, mounting height, and offset distances.

Each portable light tower in the study was used only in certain orientations. The 4000-W balloon light tower was evaluated in three mounting heights and three offset distances. The 4000-W metal halide light tower was evaluated only in three mounting heights, as changing the offset distance changed the beam angle to the vertical of the luminaires. The beam angle to the vertical was maintained at 60 degrees and the angle between the beam axis and line of sight of the driver was maintained at 90 degrees, as recommended by existing guidelines (*3*). The 800-W LED light towers were used at a fixed mounting height of 15 ft as this was the highest setting provided by the manufacturer. These light towers are often mounted on paving or milling machines,

Figure 1. Portable light towers used in study.

either one or two towers at a time (*4*). Thus, the 800-W LED light tower was used in two configurations, one light tower at a time and two light towers at time, to simulate their use in real work zones.

In order to facilitate the comparison across different light tower conditions, several characteristics, such as light tower type, their respective mounting heights, offset distances, and number of light towers, were merged to form 15 discrete categorical levels of a single variable called "light tower orientation" (see Table 2). For example, the three mounting heights and three offset distances of the 4000-W balloon light tower were merged to give nine levels. The three mounting heights of the 4000-W metal halide contributed three levels. The two conditions of the number of luminaires for the 800-W balloon LED light tower contributed two levels. Finally, a control condition with no light tower (only simulated worker under no lighting) was also used, which provided one level. Overall, the "light tower orientation" variable had 15 levels.

Mounting Height. Three different mounting heights were used. They were 15, 20, and 25 ft from the surface of the roadway. Higher mounting heights could result in lower disability and discomfort glare for the drivers.

Offset Distance. Offset distance is the location of the portable light tower on the roadway. This distance was calculated from the center of the closed lane in which the simulated work zone was established. Three offset distances were used. They were 0 ft (light tower in the lane), 10 ft (light tower in the shoulder), and 20 ft (light tower off the shoulder).

Dependent Variables

Visual performance was assessed by detection and recognition distances. These two variables are defined in the following subsections. Detection distance and recognition distance can be used to evaluate how well a lighting source can help the driver in hazard identification in a work zone.

Detection Distance. Detection distance was defined as the distance at which the participant was able to detect the presence of the simulated worker in the work zone. Detection distance has been used as a measure of visual performance in nighttime roadway visibility research (*1*, *5–8*).

Recognition Distance. The simulated worker oriented in the work zone always stood facing right or left. Recognition distance is the distance at which the participant was able to recognize the direction the simulated worker was facing. Similar to detection distance, recognition distances have been previously used in nighttime roadway visibility studies as measures of visual performance (*9*, *10*).

Discomfort Glare. Discomfort glare was measured using a rating scale as shown in Table 3. This scale has been reported to produce reliable data, with smaller numbers meaning lower discomfort glare and higher numbers meaning higher discomfort glare (*11*, *12*). The scale also has a "zero" anchor for no discomfort glare. An established discomfort glare rating scale, such as the deBoer rating scale, was not used, as research has shown that they are not good predictors of driving performance (*13*). Moreover, pilot tests showed that the

Table 2. Merging of Light Tower Characteristics into a Single Categorical Variable

Light tower type	Offset distance (ft)	Mounting height (ft)	Angle to the vertical (degree)	Number of luminaires	Light tower orientation
4000-W balloon	0	15			Ball_0_15
	10				Ball_10_15
	20				Ball_20_15
	0	20			Ball_0_20
	10				Ball_10_20
	20				Ball_20_20
	0	25			Ball_0_25
	10				Ball_10_25
	20				Ball_20_25
4000-W metal halide		15	60		MH_30_15
		20			MH_30_20
		25			MH_30_25
800-W balloon LED	0	15		1	Ball_LED_1
				2	Ball_LED_2
Control condition					No light

9-point deBoer rating scale was difficult for the participants to refer to or memorize while indicating their glare ratings, given the speed (55 mph) that was used in the current study. Participants had no issues with the selected discomfort glare rating scale for this study.

Procedure

Two participants were scheduled for each experimental session. On arrival, participants reviewed and signed the informed consent form. Participants' driver's licenses were checked for validity by the experimenters. Participants then performed a basic Snellen visual acuity test. Participants were required to have at least 20/40 vision (with or without corrective lenses) to participate in the study. Participants who did not have 20/40 vision were not used for data collection.

Once participants had completed the paperwork, the experimenter read a brief overview of the driving portion of the study and answered participants' questions. Participants were then escorted to the test vehicle and orientated to the experimental vehicle. 1999 and 2000 model years Ford Explorers served as experimental vehicles for this study and were instrumented with a data acquisition system (DAS). The DAS collected kinematic data from the vehicle's controller area network (CAN) system, including vehicle speed, differential Global Positioning System (GPS) coordinates, four video images (driver's face, forward roadway, left side of roadway, and right side of roadway), audio from the driver, manual button presses, and other input from the in-vehicle experimenter. Low-beam headlamps were used during the study. The headlamps were Hella 90 mm bi-xenon projector lamps with a single 1-F capacitor-stabilized headlamp input voltage on each vehicle. These headlamps were retrofitted along with a voltage stabilizer so that the headlamps' intensities were not affected by speed of vehicles. These retrofitted headlamps also allowed the headlamps to be at the same height from the ground across both the experimental vehicles. Before every experimental session, vehicle headlamps were aimed, and the windshields were wiped clean.

Table 3. Scale Used to Measure Discomfort Glare Rating

Description	Rating
No discomfort glare	0
Glare between non-existent and noticeable	1
Glare noticeable	2
Glare between noticeable and disagreeable	3
Glare disagreeable	4
Glare between disagreeable and intolerable	5
Glare intolerable	6

Participants drove for six laps in both uphill and downhill directions on the Virginia Smart Road. Each lap involved driving through two simulated work zones at the assigned speed limit for the study (55 mph). Participants drove in the left lane when going downhill and in the right lane when going uphill. Portable light towers were located at two stations on the Smart Road. At the first station, the 800 W LED and the 4000-W metal halide light towers were located. The 4000-W balloon light tower was located at the second station. When travelling downhill, participants first encountered the light tower at first station and then the light tower at the second station. When traveling uphill, participants encountered the stations in the reverse order. All light towers were encountered in both directions. As the participants drove through the test area, they were asked to actively scan for a simulated worker located in the work area, who was dressed in retroreflective clothing along with a hard hat (see Figure 1) as recommended by the *Virginia Work Area Protection Manual* (*14*). The vertical illuminance on the simulated worker was matched across all the light tower types, heights, offset distances, and number of light towers and was set at 50 lux. To match the vertical illuminance (light incident on a vertical plane) across each light tower and orientation, the worker's location within the work area was changed but it was always within the simulated work area.

Participants indicated when they could first see the simulated worker by saying "worker" aloud. The in-vehicle experimenter then flagged the data stream with a button press. Once participants recognized the direction the simulated worker was facing, they indicated the worker's orientation by saying "right" or "left" aloud. The in-vehicle experimenter then flagged the data stream with another button press. The GPS coordinates of the worker's locations were predetermined. The GPS coordinates at detection and recognition were cleaned up at a later point. The detections were adjusted to the point in time at which the participants said "worker" and stated the orientation, thereby eliminating the time delay due to experimenter input. A high precision GPS unit was used in the experimental vehicle and was also used to collect the GPS locations of the worker. The GPS system had an accuracy of about 0.1 m (0.33 ft).

Once the first participant vehicle was clear of the test area, the in-vehicle experimenter notified the second participant vehicle via radio that they were clear to proceed and asked the participant to provide a discomfort glare rating for the condition that they had just observed. A copy of the scale was provided for the participant for reference. After encountering the two simulated work zones in the downhill direction, the first vehicle was then parked in a turnaround and waited for the second vehicle. Once the second vehicle arrived at the turnaround, the process was repeated driving in the uphill direction. The two vehicles continued in this fashion until all light tower conditions had been observed.

Analyses

Three separate linear-mixed-model analyses were used to assess the effect of light tower orientation on detection distance, orientation recognition distance, and discomfort glare rating. Age was included as a blocking factor. The level of significance was $p < 0.05$ for all statistical tests. Where relevant, post hoc analyses (pairwise comparisons) were performed using Tukey's honest significant difference for main effects and simple effects testing for interaction effects.

Results

Detection Distance Analysis

The main effect of light tower orientation was significant, F (14, 342) = 16.06, $p < 0.0001$. The effect of light tower orientation on detection distance is shown in Figure 2.

The longest detection distances were attained under 4000-W balloon light tower at an offset of 20 ft and a mounting height of 25 ft ($M = 549.3$ m, $SD = 205.7$ m). The shortest detection distances were attained under the two-800-W balloon LED light tower orientation ($M = 253.4$ m, $SD = 75.7$ m). For the 4000-W balloon light tower, the detection distances were significantly longer for the offset distances between 0 and 20 ft at mounting heights of 20 and 25 ft only. The detection distances between the 800-W balloon LED light tower and the 4000-W balloon light tower were significantly longer at every offset distance and mounting height except at the offset distance of 0 ft and a mounting height of 15 ft. The differences in the detection distance between the 4000-W metal halide light tower and the 4000-W balloon light tower were only significant at mounting heights of 20 and 25 ft for the offset distance of 0 ft and at all mounting heights at an offset distance of 20 ft. The detection distance in the control condition was significantly lower than all light tower conditions except those of the 800-W balloon LED and the 4000-W metal halide at a height of 15 ft.

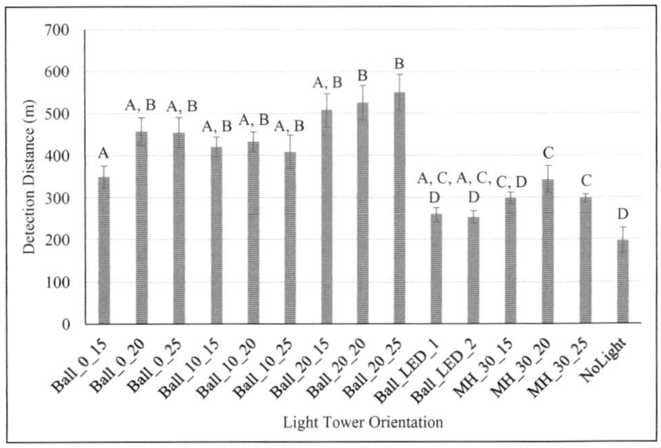

Figure 2. Effect of light tower orientation on detection distance. Values are means of detection distances and error bars reflect standard errors. Uppercase letters indicate significant ($p < 0.05$) post hoc groupings (from pairwise comparisons).

Recognition Distance Analysis

The main effect of light tower orientation on orientation recognition distance was significant, F (14, 333) = 3.45, $p < 0.0001$. The effect of light tower orientation on recognition distance is shown in Figure 3. None of the recognition distances between the light tower orientations was significant. Only the recognition distances for the control condition were significantly different from the 4000-W balloon light tower at all mounting height and offset distances except at the offset distance of 0 ft and a mounting height of 15 ft.

Discomfort Glare Rating Analysis

The main effect of light tower orientation was significant, F (14, 342) = 25.83, $p < 0.0001$. The two-way interaction between age and light tower orientation was also significant, F (14, 342) = 4.16, $p < 0.0001$. The effect of light tower orientation on discomfort glare rating is shown in Figure 4. The two-800-W balloon light tower configuration had the highest

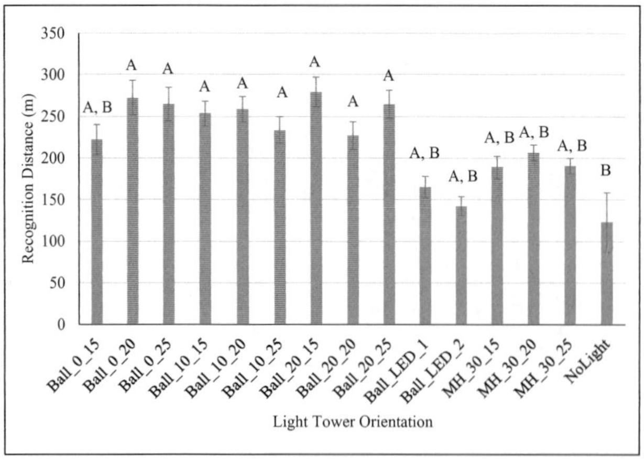

Figure 3. Effect of light tower orientation on recognition distance. Values are means of recognition distances and error bars reflect standard errors. Uppercase letters indicate significant ($p < 0.05$) post hoc groupings (from pairwise comparisons).

discomfort glare rating ($M = 3.3$, $SD = 1.7$). This light tower orientation also had significantly higher glare ratings than the 4000-W metal halide light tower at a mounting height of 20 and 25 ft, respectively. The control condition had the lowest glare rating, which was significantly lower than the glare ratings in all the light tower orientations for both the age groups. The only significant age difference was for the 4000-W metal halide light tower at a mounting height of 20 ft, where the older participants ($M = 0.8$, $SD = 0.8$) had a lower glare rating than younger participants ($M = 3.2$, $SD = 1.6$).

Discussion of Results

The goals of this paper were to evaluate the effects of mounting height, offset distance, and number of light towers in the work zone, of several commercially available portable light towers, on visual performance and discomfort glare ratings. Two major findings were evident based on the results of this study. First, an increase in the offset distance and mounting height resulted in an increase in the detection distance. Second, an increase in offset distances and mounting heights resulted in lower discomfort glare ratings.

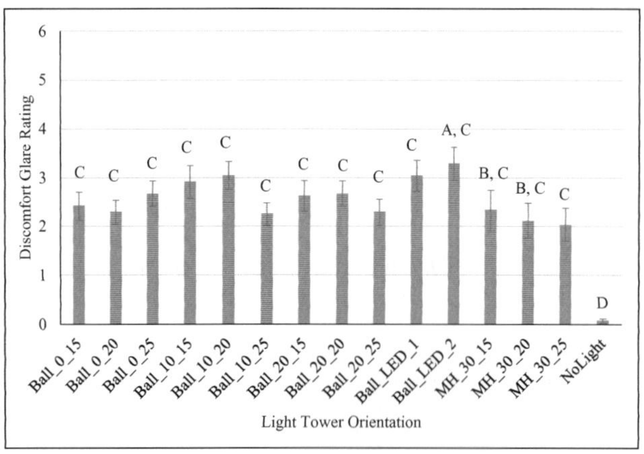

Figure 4. Effect of light tower orientation on discomfort glare rating. Values are means of discomfort glare ratings and error bars reflect standard errors. Uppercase letters indicate significant ($p < 0.05$) post hoc groupings (from pairwise comparisons).

Visual performance increased as mounting height and offset distance from the roadway increased, as evidenced by higher mean detection distances for the 4000-W balloon light tower at mounting heights of 20 ft and 25 ft for an offset distance of 20 ft. An increase in the offset distance from the roadway and the mounting height could result in lowering the veiling luminance, thereby reducing disability glare, resulting in increased visual performance. The result of higher mounting height in increasing visual performance is also supported by current recommendations on nighttime construction safety (*15*, *16*). The results from this research also show that increasing the offset distance from the roadway or locating the portable light towers off the shoulder of the roadway could also increase visual performance (by increasing detection distances), especially for the higher wattage portable light towers like the 4000-W balloon light tower, which cannot be aimed (angle of the luminaire on the light tower cannot be altered).

For light towers that could be aimed like the 4000-W metal halide light tower, an increase in the height of the light tower, while keeping the angle between the beam axis and vertical constant, could also result in increasing the visual performance (as longer but not statistically significant detection distances were observed with increase in the height of the light tower). Overall, detection distances for the 4000-W balloon light tower at highest offset and mounting height were significantly longer than those at the highest mounting height for the 4000-W metal halide light tower. This result indicates that when located at the right distance from the roadway and mounting height, a balloon light tower with a similar wattage and lumen output to a metal halide light tower could result in higher visual performance. The results of this study also showed that light towers located closest to the roadway and at the lowest mounting height of 15 ft had the lowest visual performance (lower detection distances) irrespective of the wattage and lumen output of the luminaires, as evidenced by the lower mean detection distances of 800-W LED balloon light towers and both the 4000-W balloon and metal halide light towers at a mounting height of 15 ft and an offset distance of 0 ft.

The discomfort glare rating analyses also showed that increasing the mounting height also resulted in lowering the discomfort glare ratings. This is evidenced by the significantly lower discomfort glare ratings of the 4000-W metal halide light tower at mounting heights of 20 and 25 ft compared with the 800-W LED balloon light tower mounted at a height of 15 ft. The result of decreasing glare rating along with increasing mounting height and increasing offset distance was also observed in the 4000-W balloon light tower but these ratings were not statistically significant. The results from the discomfort glare ratings analysis reinforced the results of the detection distance analysis for the portable light towers located closest to the roadway at the mounting height of 15 ft as these orientations had the highest glare ratings and shortest detection distance.

The number of the light towers (800-W LED) did not affect the visual performance or the discomfort glare ratings as shown by the lack of significant post hoc pairwise comparisons between the one vs. two light tower configurations. Although the discomfort glare ratings for the two-tower configuration were higher, it was not statistically significant. These results indicate that up to two 800-W light towers could be mounted on work zone equipment without significantly affecting the drivers' visual performance or discomfort glare ratings.

Interestingly, there were no differences in recognition distances across the several light tower orientations. This result could be because the recognition of orientation of the worker in the current study was dependent on visual acuity of the driver. Recognizing the orientation is dependent on the detection of finer details like the direction of face, hands or shoes which is different from detecting the presence of a worker. Since all the participants for the study had a visual acuity of at least 20/40, statistical significant differences between recognition drivers could not be detected. However, more research is required to confirm this finding.

Age of the drivers also did not significantly affect the drivers' visual performance, as evidenced by the lack of significant main. Age affected the discomfort glare ratings only for 4000-W metal halide light tower at a mounting height of 20 ft, for which younger drivers had a lower glare rating than older drivers.

The visual performance analyses also showed that in all the portable light tower conditions, detections and recognitions happened at distances beyond the range of the headlamps (greater than 100 m). However, the recognition distances in the control (no lighting) condition happened almost at the end of the range of the headlamps. The lack of lighting in the control condition could have resulted in the participants requiring the supplemental illumination provided by the headlamps.

The results of this study have several practical implications for nighttime work zones on limited access highways. The differences in the visual performance and discomfort glare ratings show that light tower orientation plays an important role in affecting a drivers' visibility in work zones. Light towers located closest to the roadway and at a mounting height of 15 ft had the lowest detection distances and highest discomfort glare ratings, irrespective of the wattage and lumen output. Thus, efforts should be made to locate the light towers away from the roadway and mount them higher than 15 ft. The results from the study also show that mounting heights of at least 20 ft can result in higher visual performance and lower discomfort glare. In addition to increasing the mounting height, increasing the offset distance of the portable light tower from the roadway, especially for the higher wattage (4000-W or greater) balloon light towers, can result in higher visual performance and lower discomfort glare ratings. Finally, the results also show that balloon light towers could offer better visual performance than conventional metal halide light towers of similar wattage and light output, only if they are mounted at a specified mounting height and offset distance from the roadway.

This work has a few limitations. First, there was only one worker in the work zone and no other equipment with flashing beacons was present. Second, there was no other traffic in the test area other than the experimental vehicles. These simplifications were made in the experimental design to eliminate the confounding effects that could arise due to the presence of more workers, vehicles, and traffic. Adding more workers and vehicles could potentially reduce the detection and recognition distances as drivers have to scan the work zone to perform the detection task. These results represent drivers' visual performance and glare ratings under optimal conditions and performance decrements should be expected in real road conditions. Finally, these results are only applicable to work zones on limited access highways where there are no other sources of roadway lighting other than the portable light towers. Presence of roadway lighting could further increase the detection and recognition distances and reduce the perceptions of glare (due to increase in the adaptation level of the drivers). To address the above-mentioned limitations, future work should evaluate the effects of work zone equipment, traffic density, presence of roadway lighting, and other more complex scenarios to better understand driver visual performance and their perceptions of glare in work zones.

Conclusion

In conclusion, the orientation of portable light towers, more specifically, the mounting height and offset distance, play a critical role in affecting driver visual performance and discomfort glare. Portable light towers, irrespective of wattage and lumen output, at less than 20 ft mounting height and closer to the roadway result in decreasing driver visual performance and increasing their discomfort glare. Based on the results of the study, portable light towers that cannot be aimed, like the balloon light towers, should be mounted at higher than 20 ft and at distances greater than 10 ft from the roadway. To increase visual performance, portable light towers, like the conventional metal halide light towers that can be aimed, should be mounted at a height of at least 20 ft, with the angle between the beam axis and driver line of sight greater than or equal to 90 degrees, and the angle between the beam axis and vertical less than or equal to 60 degrees.

Acknowledgments

Funding for this research was provided by the National Surface Transportation Safety Center for Excellence (NSTSCE). NSTSCE develops and disseminates advanced transportation safety techniques and innovations. The mission of the center is to use state-of-the-art facilities, including the Virginia Smart Road, to conduct the research necessary to improve driver safety in both rural and urban communities. Stakeholders of NSTSCE includethe Federal Motor Carrier Safety Administration, the Virginia Department of Transportation and the Virginia Transportation Research Council, General Motors Corporation, Travelers, and the National Safety Council. The authors would also like to thank Mr. Ken Urban (812 Illumination Inc.) for providing the necessary portable towers for evaluation.

Author Contributions

The authors confirm contribution to the paper as follows: study conception and design: Rajaram Bhagavathula, Ronald B. Gibbons; data collection: Rajaram Bhagavathula; analysis and interpretation of results: Rajaram Bhagavathula, Ronald B. Gibbons; draft manuscript preparation: Rajaram Bhagavathula. All authors reviewed the results and approved the final version of the manuscript.

References

1. Bhagavathula, R., and R. B. Gibbons. Effect of Work Zone Lighting on Drivers' Visual Performance and Perceptions of Glare. *Transportation Research Record: Journal of the Transportation Research Board*, 2017. 2617: 44–51.
2. Finley, M. D., G. L. Ullman, J. D. Miles, and M. P. Pratt. *Studies to Assess the Impact of Nighttime Work Zone Lighting on Motorists*. Texas A&M Transportation Institute, 2012.
3. Ellis, R. D., S. Amos, and A. Kumar. *NCHRP Report 498: Illumination Guidelines for Nighttime Highway Work*. Transportation Research Board of the National Academies, Washington, D.C., 2003.
4. Bhagavathula, R., R. Gibbons, A. Medina, and T. Terry. *Examination of the Current Practice of Lighting in Virginia: Nighttime Work Zones and Improving Safety through the Development of Nighttime Lighting Specifications*. Final Report VTRC 18-R3. 2017. Virginia Transportation Research Council.
5. Bhagavathula, R., and R. B. Gibbons. *Role of Expectancy, Motion and Overhead Lighting on Nighttime Visibility. Proc., Human Factors and Ergonomics Society Annual Meeting*, Vol. 57, No. 1, pp. 1963–1967, Thousands Oaks, CA, SAGE Publications, 2013.
6. Edwards, C. J., and R. B. Gibbons. Relationship of Vertical Illuminance to Pedestrian Visibility in Crosswalks. *Transportation Research Record: Journal of the Transportation Research Board*, 2008. 2056: 9–16.
7. Shinar, D. The Effects of Expectancy, Clothing Reflectance, and Detection Criterion on Nighttime Pedestrian Visibility. *Human Factors: The Journal of the Human Factors and Ergonomics Society*, Vol. 27, No. 3, 1985, pp. 327–333.
8. Zwahlen, H., and T. Schnell. Visibility of Road Markings as a Function of Age, Retroreflectivity under Low-Beam and High-Beam Illumination at Night. *Transportation Research Record: Journal of the Transportation Research Board*, 1999. 1692: 152–163.
9. Gibbons, R. B., J. Meyer, T. Terry, R. Bhagavathula, A. Lewis, M. Flanagan, and C. Connell. *Evaluation of the Impact of Spectral Power Distribution on Driver Performance*. Report No. FHWA-HRT-15-047. United States Federal Highway Administration, Washington, D.C., 2015.
10. Terry, T., and R. B. Gibbons. *Assessment of the Impact of Color Contrast in the Detection and Recognition of Objects in a Road Environment*. Virginia Tech Transportation Institute, 2011.
11. Fisher, D. Discomfort Glare in Interiors. *Presented at First International Symposium on Glare*, Orlando, Fla., 1991.
12. Tyukhova, Y. *Discomfort Glare from Small, High Luminance Light Sources in Outdoor Nighttime Environments*. The University of Nebraska-Lincoln, 2015.
13. Theeuwes, J., J. W. Alferdinck, and M. Perel. Relation between Glare and Driving Performance. *Human Factors*, Vol. 44, No. 1, 2002, pp. 95–107.
14. Virginia Department of Transportation. *Virginia Work Area Protection Manual: Standards and Guidelines for Temporary Traffic Control*. Virginia Department of Transportation, Richmond, 2011.
15. Shane, J. S., A. A. Kandil, and C. J. Schexnayder. *NCHRP Report 726: A Guidebook for Nighttime Construction: Impacts on Safety, Quality, and Productivity*. Transportation Research Board of the National Academies, Washington, D.C., 2012.
16. Ellis, R., Jr., , and S. Amos. Development of Work Zone Lighting Standards for Nighttime Highway Work. *Transportation Research Record: Journal of the Transportation Research Board*, 1996. 1529: 35–42.

The Standing Committee on Visibility (AND40) peer-reviewed this paper (18-04997).

Article

TRR
JOURNAL OF THE TRANSPORTATION RESEARCH BOARD

Transportation Research Record
2018, Vol. 2672(16) 113–121

DOI: 10.1177/0361198118774747
journals.sagepub.com/home/trr
SAGE

Safety Evaluation of Destination Lighting Treatment at Stop Controlled Cross-Intersections

Amrita Goswamy[1], Shauna Hallmark[1], Theresa Litteral[2], and Michael Pawlovich[3]

Abstract

Intersection crashes during nighttime hours may occur because of poor driver visual cognition of conflicting traffic or intersection presence. In rural areas, the only source of lighting is typically provided by vehicle headlights. Roadway lighting enhances driver recognition of intersection presence and visibility of signs and markings. Destination lighting provides some illumination for the intersection but is not intended to fully illuminate all approaches. Destination lighting has been widely used in Iowa but the effectiveness has not been well documented. This study, therefore, sought to evaluate the effect on safety of destination lighting at rural intersections. As part of an extensive data collection effort, locations with destination/street lighting were gathered with the assistance of several state agencies. After manual selection of a similar number of control intersections, propensity score matching using the caliper width technique was used to match 245 treatments with 245 control sites. Negative binomial regression was used to evaluate crash frequency data. The presence of destination lighting at stop-controlled cross-intersections generally reduced the night-to-day crash ratio by 19%. The presence of treatment or destination lighting was associated with a 33%–39% increase in daytime crashes across all models but was associated with an 18%–33% reduction in nighttime crashes. Injuries in nighttime crashes decreased by 24% and total nighttime crashes reduced by 33%. Property damage crashes were reduced by 18%.

Unlit or inadequately lit intersections reduce the ability of drivers to recognize upcoming intersections during nighttime hours. Drivers also face difficulty in safely negotiating the intersection as the lack of adequate lighting increases the likelihood of not detecting conflicting vehicles or pedestrians. In general, the nighttime crash rate is about 1.6 times that in daytime (*1*, *2*). Around 27% of total crashes and 45% of fatalities occur under dark conditions even though nighttime volume does not account for the same proportion (*3*). Hallmark et al. (*4*) conducted a cross-sectional analysis of the effects on safety of various treatments at 223 rural intersections. Their model indicated that the expected mean number of nighttime crashes was 2.01 times higher for unlighted intersections than for lighted intersections.

Standards and best practices are available for full-scale lighting of intersections. However, the addition of lighting structures and wiring can be a significant cost for rural agencies and may not be justified under lower volume conditions. One solution, commonly used in Iowa and other U.S. states, is destination lighting.

Destination lighting is intended to visually mark an upcoming intersection so drivers are alerted to its presence. In most cases, a single head is placed on the nearest utility pole which reduces the need for additional structures and wiring. The primary objective of destination lighting is to guide a driver to the intersection though this may not provide sufficient lighting to increase visibility (*5*). Destination lighting is typically placed at one approach but it is not unusual for lights to be placed on multiple approaches. Although destination lighting does not provide a full lighting footprint, it does provide some illumination for the intersection. Destination lighting can be a good countermeasure for rural intersections where

[1]Department of Civil, Construction, and Environmental Engineering, Iowa State University, Ames, IA
[2]Institute of Transportation, Iowa State University, Ames, IA
[3]South Dakota State University, Jerome J. Lohr College of Engineering, Civil & Environmental Engineering, Brookings, SD

Corresponding Author:
Address correspondence to Shauna Hallmark: shallmar@iastate.edu

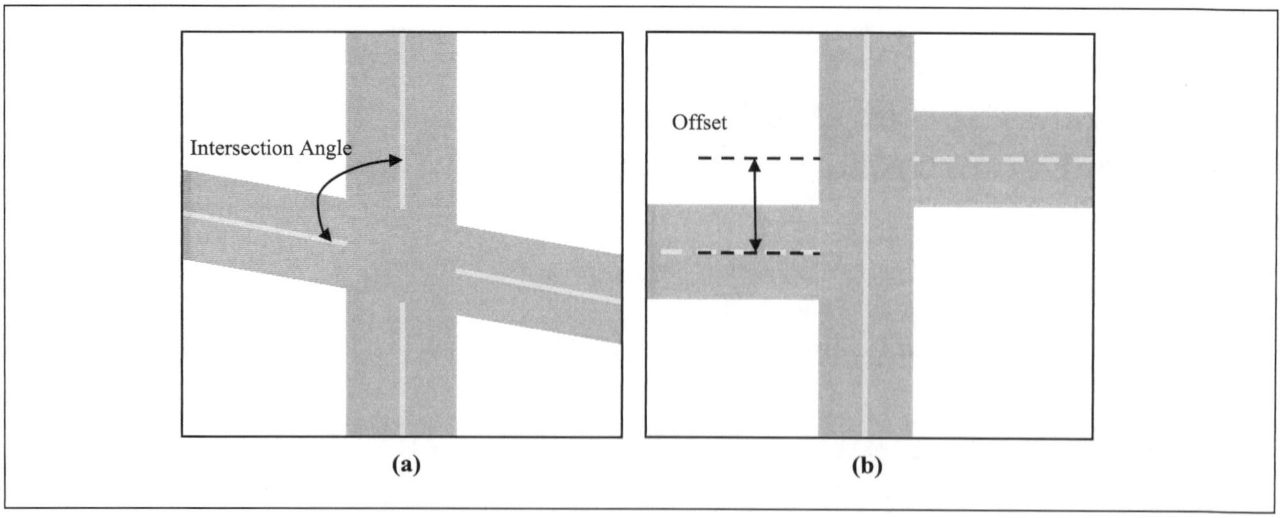

Figure 1. Intersection angle and offset distance.

nighttime crashes involve running stop sign or a failure to yield. Figure 1 exhibits a typical destination lighting configuration.

The Traffic and Safety Manual issued by Iowa Department of Transportation (DOT) provides lighting warrants for full lighting and destination lighting. According to the manual, destination lighting should be placed at intersections with more than approximately 1,750 entering vehicles in comparison to 3,500 entering vehicles for full lighting. Destination lighting should also be considered when the night-to-day crash ratio exceeds 1.0 and there exists a minimum of two reportable night crashes in a five-year period (*6*).

Although various studies have assessed the safety of intersection lighting, the effect of destination lighting has not been well studied. The objective of this research was to examine the relationship between traffic, crash, and roadway characteristics at stop-controlled cross intersections provided with destination lighting in Iowa. Crash frequency models were developed to identify the safety effectiveness of providing destination lighting at cross-intersections.

Background

Previous literature examined the safety benefits of full lighting of rural intersections that generally serve higher traffic volumes than secondary road intersections. The safety benefits of destination lighting at intersections carrying relatively low volumes have not been adequately reported. One study by Carstens and Berns (*5*) identified and evaluated safety related to destination lighting at secondary road intersections in Iowa. The study showed results from two comparisons. First, crashes were compared between before-and-after periods at 91 secondary road intersections. The study failed to show any reduction in crashes associated with the installation of destination lighting. Second, crash experience at 102 lighted intersections were compared with unlighted ones. No significant differences were observed in the crash rates at lighted and unlighted locations.

Previous studies evaluated safety of street lighting at rural intersections and indicated reduction in nighttime crashes associated with the installation. Isebrands et al. (*7*) evaluated the effectiveness of roadway lighting in reducing nighttime crashes at isolated rural intersections. The study evaluated the effect of lighting at 33 intersections with 3 years of before-and-after data. Results indicated nighttime crash rate to be a statistically significant 37% lower after lighting was installed.

In another study, Isebrands et al. (*8*) conducted a before-and-after analysis for installation of street lighting at 48 isolated rural intersections in Minnesota to evaluate the effect on crash rates. The results indicated a 13% reduction in nighttime crash frequency. The ratio of nighttime crashes to total crashes was reduced by 21% and the ratio of nighttime crashes to daytime crashes was reduced by 36% in the after period. The nighttime crash rate was seen to be reduced by 19%. On the other hand, there was an increase in the daytime crash rate by 26% from the before to after period.

Several other studies evaluated the safety of installing full scale lighting at intersections at night. Bullough et al. (*9*) examined the theoretical relationships among lighting, visibility, and safety using a statistical and an analytical approach, for different intersection types in Minnesota. Overall, the presence of roadway intersection lighting was found to be associated with an approximately 12% lower night-to-day crash ratio than at unlighted intersections. Results also suggested that visual performance improvements from roadway lighting could be used for predicting improvements in crash frequency.

Donnell et al. (*10*) evaluated the statistical association between the presence of intersection lighting and the night-to-day crash ratio in Minnesota. When traffic volume and other safety-related intersection design features were accounted for, intersections with fixed roadway lighting have fewer expected nighttime crashes and more expected daytime crashes than intersections without lighting. The nighttime crash frequency and night-to-day crash ratio decreased by 7.6% and 12% respectively. Sasidharan et al. (*11*) found that fixed roadway lighting reduces expected nighttime crashes by approximately 6%.

Bruneau and Morin (*12*) compared the safety aspects of roadway lighting at rural and near-urban three-way and four-way junctions by standard and nonstandard lighting. Rural lighting significantly reduced the nighttime crash rate by 29% for nonstandard lighting and by 39% for standard lighting, in comparison to darkness.

Walker et al. (*13*) evaluated the influence of lighting on crash frequency at 47 rural at-grade intersections in Iowa for a 3-year period before and after the installation of lighting. Installation of lighting resulted in an overall significant reduction in the average of nighttime crashes by 49%. The average nighttime crash rate per million entering vehicles was 1.89 before lighting and 0.91 after lighting indicating a reduction of 52%. Lipinski and Wortman (*14*) also analyzed crash data for rural at-grade intersections in Illinois. The results indicated that the overall night-to-day crash ratio decreased from 0.33 to 0.26 (by 22%), after the intersections were lighted. The results also showed a 45% reduction in the nighttime crash rate and a 35% reduction in the total crash rate at all intersections. Wortman et al. (*15*) collected physical characteristics, traffic volume, and crashes at rural at-grade intersections in Illinois. It was found that the lighting produced a significant reduction in nighttime crashes by 30%. The above studies are summarized in Table 1.

In many of the above studies, type of lighting (i.e., regular versus destination) was not indicated. As a result, although lighting in general has shown a positive safety benefit, the impact of destination lighting is still relatively unknown. Since destination lighting is significantly cheaper for rural agencies to install, this study

Table 1. Crash Reduction Percentages from Previous Studies

Lighting type	Study	Sites	Crash type	Change in crashes
Destination	Carstens and Berns, (*5*)	Before-after comparison of 91 secondary road intersections and comparison of 102 lighted with 102 unlighted intersections.	Total, nighttime and daytime crash rates	No reduction or differences in crashes observed
Intersection	Isebrands et al., (*7*)	33 intersections with 3 years of before-and-after data	Nighttime crashes	−37%
	Isebrands et al., (*8*)	48 rural intersections	Nighttime crashes per intersection	−13%
			Ratio of night to total crashes	−21%
			Ratio of night-to-day crashes	−36%
	Preston and Schoenecker, 1999 (*16*)	A before-after analysis of 12 intersections	Nighttime crash rate	−40%
	Bullough et al., (*9*)	Intersection in Minnesota	Night-to-day crash ratio	−12%
	Donnell et al. (*10*)	Intersection in Minnesota	Nighttime crash frequency	−7.6%
			Night-to-day crash ratio	−12%
	Bruneau and Morin, (*12*)	376 sites	All	−29%/−39%
			Property-damage-only (PDO) crashes	−37%/−48%
	Walker et al., (*13*)	47 rural at-grade intersections in Iowa	Average overall nighttime crashes	−49%
	Lipinski and Wortman (*14*)	Rural at-grade intersections in Illinois	Overall night-to-day crash ratio	−22%
			Nighttime crash rate	−45%
			Total crash rate	−35%
	Wortman et al. (*15*)	Rural at-grade intersections in Illinois	Nighttime crashes	−30%

investigated the safety effectiveness of the installation of destination lighting at rural secondary roads.

Methodology

A database was developed which included the traffic and geometric characteristics of the intersections of interest. The traffic, roadway, and geometric data of each intersection were merged with crash frequency data over a 10-year analysis period from 2005 to 2014.

Identification of Treatment Sites

The Institute of Transportation (InTrans) and the Iowa Department of Transportation (DOT) developed an intersection database for the State of Iowa. The intersection dataset also contained information about all rural and urban intersection characteristics in Iowa in the form of geographic information system (GIS) shape files.

Intersection locations on most public roadways were identified and associated characteristics such as type of traffic control (i.e., stop, yield, and signal), presence of overhead lighting, and so forth were coded. Intersections were spatially located and queried in ArcMap so that intersections with characteristics of interest could be identified. As part of the extensive process of data collection, several state agencies were contacted to identify locations with destination lighting. At the end of the process, a total of 309 intersections with destination lighting were obtained with 306 being four-legged and three being T-intersections. Given the extremely small number of T-intersections, the study was concentrated on four-legged intersections only.

Identified locations of destination lighting were checked against sources such as Google Street View, Microsoft Birdseye, and Google Aerial View to confirm the presence of the lighting. Intersection characteristics such as number of approaches and intersection angle were also extracted. Characteristics, specific to individual approaches such as roadway surface type (i.e., gravel, paved), advanced stop sign rumble strips, and advance signing were recorded. The presence of destination lighting and other intersection or approach characteristics were confirmed using multiple data sources.

Identification of Control Sites

Control sites with no lighting but with similar roadway characteristics were identified using a similar methodology as described above. One control location was selected for each treatment intersection. The first step in identifying candidates was to check intersections in proximity to the treatment intersection. A control intersection was manually identified using ArcGIS to match a corresponding treatment intersection using the following important variables (in most cases the first four criteria were matched for treatment and control intersections):

- Type of intersection: intersection configuration (i.e., four-lane/two-lane stop control; two-lane/two-lane stop control)
- Type of traffic control (i.e., two-way stop)
- Proximity
- Volume on minor vs. major roads (within 500 vehicles per day if possible)
- Channelization (turn lane)
- Presence of rumble strips or countermeasures (such as raised median)

Intersection Approach Database

Once an initial set of locations was identified, additional data for individual approaches of the treatment and control approaches were aggregated. The intersection approach dataset contained the following factors:

- Traffic mode
- Pavement surface type
- One- or two-way
- Intersection angle
- Offset distance
- Left-turn type
- Left-turn bay length
- Number of exclusive left-turn lanes
- Left-turn offset
- Right-turn channelization
- Number of exclusive right-turn lanes
- Traffic control exclusive right-turn
- Right-turn offset
- Median type
- Traffic control
- Crosswalk type
- Transverse rumble strips

Intersection angle was recorded as the interior (smaller angle) of the most angled member of the intersection. It reflects the angle of traffic as it approaches the intersection. The offset distance was the measure of distance from the centerline of one roadway to the centerline of the opposite roadway. Both of these attributes were measured in ArcGIS. Figure 1 shows the measurement of the intersection angle and offset distance.

Traffic Data

The Iowa DOT maintains a roadway inventory used to obtain the traffic volume data for each of the approach roadways in an intersection. Annual Average Daily

Traffic (AADT) figures for each of the approaches were obtained for the middle year for the ten-year study period 2006–2016. Daily entering vehicles were calculated for all the intersections using Equation 1 which assumes a 50/50 directional split of segment volumes (*6*).

$$\text{DEV} = \frac{(AADT_N + AADT_S + AADT_E + AADT_W)}{2} \quad (1)$$

where:

DEV = Daily entering volume for an intersection
$AADT_N$ = AADT from north approach
$AADT_S$ = AADT from south approach
$AADT_E$ = AADT from east approach
$AADT_W$ = AADT from west approach

Iowa Crash Database

The Iowa DOT maintains a historical crash database that includes geographically referenced information detailing the driver, vehicle, roadway, and environmental factors associated with each police-reported crash. Crash-level data provided information about the location, date, severity level, and other general information about each crash. Crashes occurring within 250 feet of each intersection were obtained for the 11 years from 2006 to 2016. Various records were consulted to ensure lighting was present at treatment sites during the entire 11-year study period.

Crash Severity levels were designated as follows: Fatal Injury (K), Disabling Injury (A), Visible Injury (B), Possible Injury (C), and Property-Damage-Only (PDO). Time of crash was used to identify nighttime crashes. Crashes at dusk and dawn were not included in the calculation of nighttime crashes.

Total nighttime and daytime crashes were determined for each of the treatment and control sites. An original model was developed to develop a CMF for nighttime crashes. It was not possible to determine when lighting had been installed so a cross-sectional rather than a before-and-after analysis was conducted. However, results were not conclusive. It was determined that this was likely because lighting had been installed at locations which had a nighttime crash problem which differed from control intersections and thus resulted in selection bias.

A preliminary simple comparison of the data indicated that there were significant differences in the ratio of day-to-night crashes at treatment versus control intersections suggesting the treatment is more effective at night. This method has been used by others to compare the relative safety of lighting such as Jackett and Frith (*17*) and Bhagavathula et al. (*18*) who used the nighttime to day-time crash ratio to assess road luminance.

Table 2. Night-to-Day Crash Ratios

	Total (KABCO) observed daytime crashes (D)	Total (KABCO) observed nighttime crashes (N)	N/D
Control	329	126	0.38
Treatment	462	87	0.19

As noted in Table 2, the night-to-day crash ratio for treatment sites with destination lighting was 0.19 while the ratio for control sites was 0.38. As a result, this simple comparison indicates that the ratio of night-to-day crashes at treatment sites is about 50% lower than for control sites.

Descriptive Statistics

Table 3 summarizes the descriptive statistics of variables for treatment and control intersections separately. The 245 treatment and 245 control intersections experienced a total of 213 nighttime crashes during the analysis period. Table 3 also provides mean and standard deviation of the variables. For the continuous variables, DEV, it can be seen that the mean values are comparable for both the treatment and control sites. For categorical variables the mean shows the percentage of intersections in which the variable was present. For example, one exclusive right-turn lane was present in 17% and 7% of the treatment and control intersections respectively.

Statistical Analysis

Estimation of Propensity Scores

Propensity score matching was used to compare the "likeness" of treatment and control sites. The method matches a treatment case with one or more control cases based on each case's propensity score. This procedure proves to be helpful for observational cross-sectional studies. It has been pointed out by Cao et al. (*19*) that an almost "identical" observation in the control group could be found for an observation in the treatment group. This matching is approximately equivalent to the process in which one of the two "same" observations is assigned into a treatment group and the other is assigned into a control group. If the process is repeated for all observations in the treatment group, observations in the matched treatment group should not differ from those in the matched control group.

Propensity score matching was performed to find balance between treatment and control intersections utilized in the study. A caliper was used to avoid large differences between scores as suggested by Guo and Fraser (*20*),

Table 3. Descriptive Statistics for the Control and Treatment Data

Variables	Treatment intersections (with destination lighting)		Control intersections (without destination lighting)	
	Mean	Std. dev	Mean	Std. dev
Stop-line rumble strip present	0.02	0.15	0.02	0.13
Paved major paved minor road	0.38	0.49	0.22	0.41
Paved major gravel minor road	0.62	0.49	0.78	0.41
Lanes: two minor three/four major road	0.03	0.15	0.03	0.14
Lanes: two minor two major road	0.97	0.17	0.97	0.18
One exclusive left turn present	0.02	0.14	0.02	0.15
Right channelization present	0.04	0.19	0.02	0.15
One exclusive right turn present	0.17	0.38	0.07	0.26
Daily entering vehicles (DEV)	1548.15	1268.13	1587.75	1409.8
Intersection angle (°)	88.74	5.41	89.02	4.77
Offset distance	0.02	0.26	0.10	0.83
All-way stop	0.02	0.14	0.02	0.14
One-way stop controlled	0.07	0.25	0.10	0.29
Two-way stop controlled	0.91	0.25	0.88	0.33

Sasidharan and Donnell (*11*). A caliper width of 0.25 times the standard deviation of the propensity scores for the treated locations was used in previous research (*12*). A range of caliper widths was tested by Wood et al. (*21*). It has been pointed out in the same study that narrow caliper widths tend to reduce the sample size for use in the potential outcomes model, relative to a wider caliper width. Different caliper widths of (0.25/0.20/0.15/0.10) times standard deviations were used and a caliper was set at 0.2 times the standard deviation beyond which improvement in matching was not observed.

The dependent variable was a dummy variable that indicated the presence or absence of destination lighting. The propensity score was estimated by using a logit model, which has the following form (*21*):

$$\Pr(y_i = 0) = \left\{1 + \exp\left[-(\beta_0 + \beta_j X_{ij})\right]\right\}^{-1} \quad (2)$$

where,

$\Pr(y_i = 0)$ = propensity score (probability of having destination lighting),

β_0 = intercept term

β_j's = parameters to be estimated, and

X_{ij}'s = independent variables.

This logit model form was applied to different selections from the list of available covariates, and the model with the best fit was used to estimate the propensity scores. The best fit represented the set of covariates that produced significant coefficients in the estimation of the logit model and that minimized the negative log-likelihood of the model.

The "MatchIt" package within the statistical software 'R' was used to generate the propensity scores for the project. The data set was set up in a way such that it consisted of cases in rows and variables in columns. A nearest-neighbor technique was used to match the cases. This technique matched a treated unit to a control unit that is closest in distance measure. The propensity score method selected for this study was caliper-based (0.2 times standard deviation), 1:1 nearest-neighbor matching without replacement (*20*). About 245 treatment and control sites were matched according to variables of traffic volume, presence of rumble strips, county, and so forth.

Crash Frequency Analysis

Installation dates could not be obtained for more than a small percentage of treatment sites. As a result, a before-and-after study could not be conducted as noted above. Models were developed for target intersection-related crashes. A target crash was defined as any non-animal, intersection-related crash. Animal crashes were not included since the presence of an animal is an unexpected event and lighting has not historically been used to address animal crashes.

Separate models were developed across all severities (fatal, major, injury, minor, possible, and property-damage-only), for injury severities (fatal, major, injury, minor) and property-damage-only crashes. Cross-sectional crash models using negative binomial generalized linear regression analysis were developed with an indicator variable for the presence and absence of destination lighting. Once the databases were compiled, a series of statistical analyses were conducted to ascertain how crashes were affected by traffic, geometric, and site characteristics. Various count data model frameworks were considered in the development of crash prediction models. As a starting point, a Poisson regression model was estimated, where the probability of intersection *i* experiencing y_i crashes during a specific period is:

$$P(y_i) = \frac{\exp(-\lambda_i)\lambda_i^{y_i}}{y_i!} \quad (3)$$

where λ_i is the Poisson parameter for segment *i*, which is equal to the expected number of crashes at the intersection during the analysis period, $E[y_i]$. Poisson regression models are estimated by specifying Poisson parameter λ_i as a function of explanatory variables, with the most common form given below:

$$\lambda_i = EXP(\beta X_i) \quad (4)$$

where X_i is a vector of explanatory variables and β is a vector of estimable parameters. The negative binomial model is derived by rewriting the Poisson parameter for each segment *i* as follows:

$$\lambda_i = EXP(\beta X_i + \varepsilon_i) \quad (5)$$

where $EXP(\varepsilon_i)$ is a gamma-distributed error term with mean one and variance α. The term α is known as the over-dispersion parameter, which reflects the additional variation in crash counts beyond the Poisson model (where α is assumed to be equal to zero). The addition of this term allows the variance to differ from the mean as:

$$VAR[y_i] = E[y_i] + \alpha E[y_i]^2 \quad (6)$$

Elasticities were calculated according to Donnell et al. (*10*). Elasticities can be defined as the percentage change in expected crash frequency given a 1% change in a continuous explanatory variable. For log-log functional form, it is simply calculated as the parameter estimate. Pseudo-elasticities were calculated for indicator variables. Pseudo-elasticities represent the percentage change in expected crash frequency given a change in the value of the indicator variable from zero to one. The elasticity of the expected crash frequency for an indicator variable can be defined as Equation 7:

$$\text{Pseudo\_elasticities} = \exp(\beta) - 1 \quad (7)$$

Results

Donnel et al. (*10*) also used negative binomial regression to model night and day crash ratios. Elasticities for continuous variables like AADT and pseudo elasticities in the form of percentages for indicator variables in the model were calculated according to a study by Donnell et al. (*9*). The elasticities computed based on the daytime and nighttime crash frequency models shown in Table 4 are provided in Table 5.

Separate models for nighttime and daytime crashes were evaluated. The parameter estimates of the statistically significant variables (at 95% confidence level) in the cross-sectional models are shown in Table 4. The standard errors and p-values are also provided for the

Table 4. Nighttime and Daytime Crash Frequency Models

		Nighttime crashes			Daytime crashes		
Crash type	Variables	Parameter estimates	Std. error	*p*-value	Parameter estimates	Std. error	*p*-value
KABCO (Total crashes)	(Intercept)	−7.441	0.775	0.000	−7.404	0.499	0.000
	Destination lighting (treatment)	−0.406	0.154	0.008	0.306	0.101	0.002
	LnDEV	0.960	0.103	0.000	1.035	0.066	0.000
	Paved major paved minor road	0.337	0.157	0.032	0.651	0.101	0.000
	One exclusive left turn present	−0.769	0.431	0.074	NA	NA	NA
	All-way stop	−1.327	1.061	0.211	−0.408	0.521	0.434
	Two-way stop controlled	−0.399	0.233	0.086	−0.243	0.160	0.128
	Alpha*	0.206	0.134	0.158	0.322	0.062	0.000
KABC (Injury crashes)	(Intercept)	−7.208	1.524	0.000	−7.273	0.528	0.000
	Destination lighting (treatment)	−0.277	0.341	0.417	0.336	0.136	0.013
	LnDEV	0.777	0.211	0.000	0.879	0.082	0.000
	Paved major paved minor road	0.403	0.374	0.282	0.792	0.136	0.000
	Alpha*	4.693	1.796	0.009	0.400	0.114	0.000
PDO (Property-damage-only crashes)	(Intercept)	−2.500	0.469	0.000	−9.264	0.785	0.000
	Destination lighting (treatment)	−0.209	0.101	0.039	0.287	0.135	0.034
	LnDEV	0.357	0.063	0.000	1.172	0.101	0.000
	Paved major paved minor road	NA	NA	NA	0.538	0.144	0.000
	One exclusive right turn present	−0.169	0.134	0.207	NA	NA	NA
	Alpha*	0.226	0.001	0.979	0.495	0.124	0.000

*Over-dispersion factor.

Table 5. Elasticities for Daytime and Nighttime Crash Frequencies

Crash type	Variables	Nighttime crashes		Daytime crashes	
		Parameter estimates	Elasticity or pseudo-elasticity	Parameter estimates	Elasticity or pseudo-elasticity
KABCO (Total crashes)	Destination lighting (treatment)	−0.406	−33.369	0.306	35.798
	LnDEV	0.960	0.960	1.035	1.035
	Paved major paved minor road	0.337	40.074	0.651	91.746
	One exclusive left turn present	−0.769	−53.652	NA	NA
	All-way stop	−1.327	−73.473	−0.408	−33.502
	Two-way stop controlled	−0.399	−32.901	−0.243	−21.573
KABC (Injury crashes)	Destination lighting (treatment)	−0.277	−24.195	0.336	39.934
	LnDEV	0.777	0.777	0.879	0.879
	Paved major paved minor road	0.403	49.631	0.792	120.781
PDO (Property-damage-only)	Destination lighting (treatment)	−0.209	−18.860	0.287	33.242
	LnDEV	0.357	0.357	1.172	1.172
	Paved major paved minor road	NA	NA	0.538	71.258
	One exclusive right turn present	−0.169	−15.549	NA	NA

parameter estimates or, in other words, beta estimates in Table 4. It can be seen from Table 4 that the coefficient of destination lighting is negatively related to all types of nighttime crashes which suggests that destination lighting has the potential to reduce nighttime crashes.

The elasticities for daily entering volumes are similar across all the nighttime and daytime crash frequency models. In both cases, however, the nighttime elasticities are lower than the daytime elasticities. A 1% increase in major road average AADT was associated with approximately a 0.4% to 1.2% increase in crash frequency. It was seen that the treatment elasticities change sign when the expected daytime and nighttime crash frequencies across all models are compared. This suggests that the presence of treatment or destination lighting is associated with a 33%–39% increase in daytime crashes across all models, but was associated with an 18%–33% reduction in nighttime crashes.

Injury nighttime crashes decreased by 24% and total nighttime crashes by 33%. Property-damage-only crashes were reduced by 18%. The correlation between increased daytime crashes and destination likely suggests the treatment was applied at locations with a higher number of crashes.

The presence of an exclusive left-turn lane reduced total nighttime crashes by 54%. All-way stop and two-way stop controlled intersections were safer than the one-way stop controlled intersections for both daytime and nighttime total crashes. Since only four-approach intersections were included, a one-way stop controlled intersection was the rare situation where only one of four approaches had stop sign control. Intersections with paved major as well as paved minor roads did not show any safety effect when compared with intersections with paved major and unpaved minor roads across all the models. The presence of an exclusive right-turn lane reduced property-damage-only nighttime crashes by 16%.

Discussion

This study examined the safety effectiveness of the installation of destination light at stop controlled cross intersections in Iowa. Preliminary night-to-day crash ratios showed the night-to-day crash ratio for treatment sites with destination lighting was 0.19 while the ratio for control sites was 0.38. This indicated that around 0.19 nighttime crashes result for every daytime crash at treatment sites while around 0.38 night crashes occur at control sites for every daytime crash. Thus, this simple comparison showed that the ratio of night-to-day crashes at treatment sites is about 50% lower than for control sites.

Various factors affecting the frequency of crashes occurring on the study sites were also analyzed. Propensity score matching using the caliper width technique was used to match treatment and control sites in this study. Crash frequency data were analyzed using negative binomial regression models. The presence of treatment or destination lighting was associated with a 33%–39% increase in daytime crashes across all models, but was associated with an 18%–33% reduction in nighttime crashes. This suggests, as expected, that destination lighting has been utilized at locations which exhibited a crash problem. As a result, destination lighting provided a significant safety benefit. Injury nighttime crashes decreased by 24% and total nighttime crashes reduced by 33%. Property damage crashes were reduced by 18%.

Ultimately, the results of this study provided important insights that may be used for understanding the importance of providing destination lighting at stop controlled cross intersections. Moving forward, the extensive databases developed as a part of this study may also be supplemented with additional information, such as shoulder information and information of guard rails. As intersections continue to be an emphasis area for improving safety, the identification of these factors will allow for the proactive and cost-effective implementation of various safety measures.

Acknowledgments

The team wishes to thank the Iowa Department of Transportation and the Midwest Transportation Center who provided funding for this project.

Author Contributions

The authors confirm contribution to the paper as follows: study conception and design: Hallmark, Goswamy, Pawlovich; data collection: Litteral, Goswamy; analysis and interpretation of results: Hallmark, Goswamy, Pawlovich; draft manuscript preparation: Hallmark, Goswamy, Pawlovich, Litteral. All authors reviewed the results and approved the final version of the manuscript.

References

1. Hasson, P., and P. Lutkevich. Roadway Lighting Revisited. *Public Roads*, Vol. 65, No. 6, 2002.
2. Opiela, K. S., C. K. Andersen, and G. Schertz. Driving After Dark. *Public Roads*, Vol. 66 No. 4 2003.
3. NHTSA. *Traffic Safety Facts 2003*. US Department of Transportation and National, Highway Traffic Safety Administration, 2003. https://crashstats.nhtsa.dot.gov/Api/Public/Publication/809767
4. Hallmark, S., N. R. Hawkins, O. Smadi, C. Kinsenbaw, M. Orellana, Z. Hans, and H. N. Isebrands. *Strategies to Address Nighttime Crashes at Rural, Unsignalized Intersections*. Iowa DOT/Iowa State Univ., Center for Transportation Research and Education, Ames, 2008. http://www.intrans.iastate.edu/research/detail.cfm?projectID_1113252604. Accessed November 19, 2009.
5. Carstens, R. L., and L. D. Berns. *Roadway Lighting on Secondary Roads in Iowa*. Iowa Department of Transportation, 1984. http://publications.iowa.gov/16235/1/IADOT_hr251_Road_Lighting_Second_Roads_Iowa_1984.pdf
6. Iowa DOT. *Traffic and Safety Manual. Iowa Department of Transportation*. Office of Traffic and Safety, Ames, Iowa Accessed July 2017, 2004. https://www.iowadot.gov/traffic/manuals/tsmanual.aspx
7. Isebrands, H., S. Hallmark, W. Li, T. McDonald, H. Preston, and R. Storm. Roadway Lighting Shows Safety Benefits at Rural Intersections. *ASCE Journal of Transportation Engineering*, Vol. 36, No. 11, 2010, pp. 949–955.
8. Isebrands, H., S. Hallmark, Z. Hans, T. McDonald, H. Preston, and R. Storm. *Safety Impacts of Street Lighting at Isolated Rural Intersections - Part II*. Minnesota Department of Transportation, St. Paul Minnesota, 2006.
9. Bullough, J. D., E. T. Donnell, and M. S. Rea. To illuminate or not to illuminate: Roadway lighting as it affects traffic safety at intersections. *Accident Analysis and Prevention*, Vol. 53, 2013, pp. 65–77.
10. Donnell, E. T., R. J. Porter, and V. N. Shankar. A Framework for Estimating the Safety Effects of Roadway Lighting at Intersections. *Safety Science*. Vol. 48. 2010. pp. 1436–1444.
11. Sasidharan, L., and E. T. Donnell. Application of Propensity Scores and Potential Outcomes to Estimate Effectiveness of Traffic Safety Countermeasures: Exploratory Analysis Using Intersection Lighting Data. *Accident Analysis and Prevention*, Vol. 50, 2013, pp. 539–553.
12. Bruneau, J. F., and D. Morin. Standard and Nonstandard Roadway Lighting Compared with Darkness at Rural Intersections. *Transportation Research Record: Journal of the Transportation Research Board*, 2005. 1918: 116–122.
13. Walker, F. W., and S. E. Roberts. Influence of Lighting on Accident Frequency at Highway Intersections. *Transportation Research Record: Journal of the Transportation Research Board*, 1976. 562: 73–77.
14. Lipinski, M. E., and R. H. Wortman. Effect of Illumination on Rural At-Grade Intersection Accidents. *Transportation Research Record: Journal of the Transportation Research Board*, 1978. 611: 25–27.
15. Wortman, R. H., and M. E. Lipinski. *Rural At-grade Intersection Illumination*. Illinois Department of Transportation, Illinois. 1974.
16. Preston, H., and T. Schoenecker. *Safety Impacts of Street Lighting at Isolated Rural Intersections*. Final Report MN/RC-1999-17. Minnesota Department of Transportation, April 1999.
17. Jackett, M., and W. Frith. Quantifying the Impact of Road Lighting on Road Safety — A New Zealand Study. *International Association of Traffic and Safety Sciences (IATSS) Research*, Vol. 36. 2013. pp. 139–145.
18. Bhagavathula, R., R. Gibbons, and C. Edwards. Relationship between Roadway Illumination Level and Nighttime Rural Intersection Safety. *Transportation Research Record: Journal of the Transportation Research Board*, 2015. 2485: 8–15.
19. Cao, J., and J. Schoner. *Transportation Impact of Transitways: A Case Study of Hiawatha March Light Rail Transit in Minneapolis*. Transitway Impacts Research Program University of Minnesota. Minneapolis, Minnesota, 2013.
20. Guo, S., and M. W. Fraser. *Propensity Score Analysis*. Sage Publications Inc., Thousand Oaks, CA, 2010.
21. Wood, S. J., E. T. Donnell, and R. J. Porter. Comparison of Safety Effect Estimates Obtained from Empirical Bayes Before–after Study, Propensity Scores-potential Outcomes Framework, and Regression Model with Cross-sectional Data. *Accident Analysis and Prevention*. Vol. 75, 2015, pp. 144–154.

The Standing Committee on Visibility (AND40) peer-reviewed this paper (18-05605).